Process plant commissioning

A user guide

Second edition

Process plant commissioning

A user guide

Second edition

Edited by David Horsley

IChem**E**
INSTITUTION OF CHEMICAL ENGINEERS

Published by
Institution of Chemical Engineers,
Davis Building,
165–189 Railway Terrace,
Rugby, Warwickshire CV21 3HQ, UK

© 1998 Institution of Chemical Engineers
A Registered Charity
First Edition 1990
Second Edition 1998

ISBN 0 85295 398 4

Cover illustration adapted from a photograph by Krupp Uhde, Dortmund, Germany

Printed in the United Kingdom by Bookcraft Ltd, Bath.

Foreword to the second edition

The Institution's Engineering Practice Committee is to be warmly commended for commissioning the updating of this excellent user guide.

Successful project execution demands the highest professionalism from all members of the project team. Many current projects are 'commissioning driven' as a means of reducing the overall project schedule. On these projects, all the preceding activities — that is, engineering, procurement and, indeed, construction itself — are directed not just towards the construction or mechanical completion of the plant, but also through the specific commissioning sequences required to overall final acceptance.

All appropriate contractual arrangements must identify and facilitate activities that cross the complex interfaces between construction, precommissioning and commissioning. The additions to Chapter 2 are welcomed, in particular the emphasis placed on ensuring good scope definition and the establishment of clear responsibilities, so that appropriate documentation and procedures can be put in place. Lack of definition in the split of work between contractors and client often causes misunderstanding and potential conflict between the various parties, at a time in the project, when, more than ever, a collaborative team effort is required.

Since the last edition, significant changes to legislation have come into force — for example, the Control of Substances Hazardous to Health (COSHH) Regulations 1994 which clearly set out the obligations of employers, as well as the Construction, Design and Management (CDM) Regulations 1994 which require that the risks of commissioning, as well as construction, are taken into account throughout the design.

Environmental legislation, following the introduction of the Environmental Protection Act 1990, has also arrived since the last edition. Such legislation requires that the engineer maintains a duty to protect the environment at all times, not least during the construction and commissioning phases. Environmental legislation is a developing field and the addition of this topic to the guide is a necessary and welcome addition.

Process plant engineering and construction is now very much an international business, and increased personal mobility is a fact of life for many engineers. They are now expected, and required, to spend a good part of their working career in foreign locations. Such locations pose particular challenges and the guide now recognizes the nature of such work and describes related issues.

A new chapter has been added on the subject of problem avoidance. The reader is recommended to question whether the critical execution issues that are appropriate to the project have been recognized and considered. Today's successful projects are those that have fully taken aboard the lessons learned from previous projects. The causes of many problems often have common roots and can be remarkably similar on otherwise very different projects. Modern engineering design tools and other integrated management systems do now play a big role in this regard. In addition, the effective use of electronic 3-D CAD modelling, using clash detection techniques, minimizes expensive rework at site and the associated derivative effect of uncontrolled modifications.

Plant handover, by process system, allows the progressive precommissioning and commissioning of the plant and offers significant schedule advantages. It does, however, require increased planning effort. In addition, a comprehensive Safe System of Work has to be set up to allow construction activities to proceed safely, whilst precommissioning and commissioning work proceeds in parallel.

The need to maintain the integrity of the design is of prime importance during construction, precommissioning and commissioning. Quality Assurance systems need to be established for these activities to protect design integrity and ensure the plant is built as per design. A useful set of appendices and example check-lists and check-sheets are contained within this guide.

Commissioning is as much a management task as a technical task. The large quantity of documentation and records to be handled must be recognized, and an appropriate database is essential to manage the voluminous amount of construction and commissioning records generated on all large projects.

The editor of this second edition is to be applauded for enhancing what was already a very useful and readable guide, and for providing clear advice on the overall subject of process plant commissioning, which is obviously a key phase of project execution.

John Blythe
Chairman and Chief Executive
Foster Wheeler Ltd
December, 1997

Foreword to the first edition

The Institution's Engineering Practice Committee is to be warmly congratulated on initiating a user guide on 'process plant commissioning', and on identifying a very strong team to execute the project in exemplary fashion. The guide they have produced is rich in valuable perceptions based on much distilled wisdom and experience, in addition to providing a systematic treatment of the subject.

By the time the commissioning stage is reached, major capital will have been spent, and the corporate climate is likely to be one of high expectancy, some anxiety and a degree of impatience. The project will have been undertaken in the expectation of meeting a market need, often with a great deal of urgency.

This adds to the psychological burden of what is in any event the climax of a substantial multidisciplinary engineering effort led by the process engineering discipline, not infrequently preceded and supported by considerable research investment.

Looking back on a career rich in commissioning experience I have no doubt that, of all the project stages, the commissioning phase is the most demanding in skill, judgement, leadership, perseverance, courage and sheer physical endurance. It also has abnormal potential for hazard.

Whereas there will be wide variations in scale and complexity, in contractual arrangements and in the degree of process innovation, commissioning imposes a specific approach to the management of the four interlocking elements involved:

- the commissioning team;
- the plant;
- the process;
- raw materials, intermediates and products.

The additional dimensions required in each case, compared with normal operation, emerge from the guide with great clarity.

It is equally explicit on the main ingredients of success:
- meticulous attention to safety and hazards;
- an ably led, well-balanced, well-trained and committed commissioning team, capable of absorbing the physical and psychological stress;
- adequate involvement in the design phase and safety studies;
- thorough planning, implementation and control of commissioning preparations;
- ready availability of help from supporting disciplines to deal with specific problems identified;
- an expeditious approval system for agreeing plant modifications;
- a well structured relationship at senior level with site construction management, and with future operational management.

I applaud the reference to the 'postcommissioning phase'. The proper recording of modifications is a vital task, as is the preservation of performance tests as datum points for continuing operation. Of even greater long term significance is a detailed postcommissioning review, recording all that has been done to improve the reliability and operability of the plant during the commissioning phase. It should moreover capitalise on the deep insights gained, by defining targets for further performance improvement and cost reduction.

The work is enhanced by a most useful set of Appendices, providing a great variety of check-lists including, *inter alia*, final checks before introducing process materials, safety assessment of modifications, equipment check out by categories, piping systems, electrics, control systems and storage.

This guide makes available to the profession much hard won and valuable experience for which we are greatly indebted to the members of the Working Party, and their employing organizations. It will make a major contribution to safer and more efficient commissioning of process plants in the years ahead, and to the professional development of process engineers.

Roger Kingsley
Director, Capcis Ltd
Past President of IChemE
September 1990

Membership of the working party for the first edition

Members

J.S. Parkinson (Chairman)	Courtaulds Research
J. Broughton	Courtaulds Engineering
D.M.C. Horsley	British Nuclear Fuels plc
J. Love	University of Leeds
F.L. Owen	Lankro Chemicals Ltd
L. Pearson	ICI (retired)
R.B.S. Prior	Humphreys & Glasgow Ltd (retired)
N.F. Scilly	Health and Safety Executive
P. Snowdon	University of Leeds
L.F. Stooks	APV Baker Ltd

Corresponding members

M.H.J. Ashley	John Brown Engineers & Constructors Ltd
J.B. Brennard	Consultant
K. Griffiths	Leigh Group
J. Lindley	ICI

Acknowledgements

The support and encouragement given to members of the Working Party by their respective employing organizations during the preparation of the first edition of this guide is gratefully acknowledged.

The Editor wishes to express his gratitude to: Bob Akroyd and Giles Gillett of Kvaerner Davy, Jim Bulman, Amanda Lomax and David Witt of British Nuclear Fuels, Roger Simpson of Zeneca, David Lonsdale and James Howells of Monsanto and Professor Stephen Wearne for their help and advice in updating the guide.

Responsibility for the contents remains with the authors and the editors.

Extracts from BS 6739, 1986, are reproduced with the permission of the British Standards Institution (BSI). Complete copies can be obtained by post from BSI Sales, 389 Chiswick High Road, London, W4 4AL. Main switchboard telephone number: +44 181 996 9000, Main facsimile: +44 181 996 7400.

Contents

Introduction

1

This guide provides information on the commissioning of process plant. The guide does not restrict itself to chemical plant and is intended to be equally useful to those employed in the food and allied industries. Likewise the guide should be useful for the commissioning of small plant as well as large plant. Although most of the guide deals with the commissioning of plant in the UK, Section 10.3 (see page 60) addresses some of the problems of commissioning foreign plant.

The aim is to provide a document which gives the non-specialist engineer advice on how to set about the problem of commissioning either a new plant or indeed a modification to an existing plant. Some aspects of decommissioning process plant have also been included. It must be stressed that typical schedules, check-lists and so on are included as a guide to good practice. They are only a 'starter' and those involved in commissioning must fully understand the technology and equipment that they will be responsible for, and be masters of the systems used to plan or analyse the commissioning operation.

Careful consideration has been given to the layout of this guide. It is suggested that users start at the beginning and work through the text, although for those needing only a 'refresher' some useful check-lists have been provided in the Appendices. In addition, a glossary of terms includes a definition of 'specialist' terminology. Commissioning of process plant is most commonly the responsibility of one of the following:

• the owner (operator) if, for example, the project is managed and executed 'in-house' by an operating company utilizing perhaps its own process technology;

• a contractor, if responsibility for the project or for some aspects of it has been 'let' by the owner (client). In such cases the process package may be proprietary to the client, contractor or licensor, or to a combination of these.

Whichever case applies, the commissioning operation requires meticulous attention both in the preparation of stages and in subsequent execution. This guide seeks to identify and highlight the more important features to be considered in all phases of commissioning. Much of it is written assuming the use of a contractor to undertake commissioning. This is because the client/contractor interface brings added complexity which has to be managed effectively if programme, budget and, above all, safety in all aspects of the project are to be maintained. The advice is equally relevant for 'in-house' commissioning which generally involves subcontract or agency (hired) personnel.

Large, novel, 'made-to-order' projects require careful appraisal of the project (and hence commissioning) strategy. A highly focused, close-knit, multidisciplinary task force approach including R&D, design, construction, commissioning and operations personnel has often been used successfully. If such a project is anticipated, the reader is recommended to seek advice from people who have managed such projects since it is impossible in a guide such as this to cover all the issues and uncertainties which will have to be taken into account in such complex projects.

A block diagram in Figure 1.1 shows the phases of a typical project and the place of the commissioning activities described in this guide.

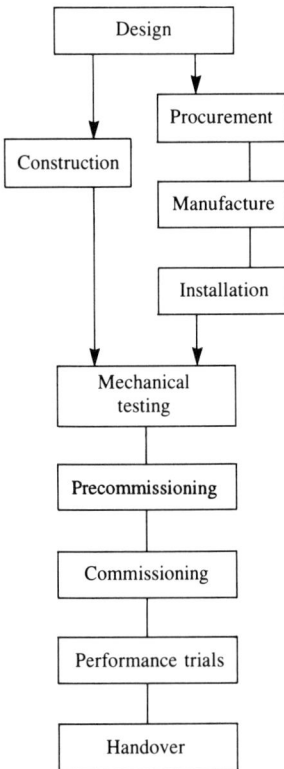

Figure 1.1 Typical phases of a project

Contracts, planning and administration

2

2.1 Engineering contracts

The responsibilities of the client and contractor throughout the project to the completion of commissioning must be clearly understood and defined in engineering contracts. Any new project may well involve technology licensed from a third party and the responsibilities of this licensor must also be clear, understood and set out in the relevant contract.

A policy for supervising contractors and subcontractors should be decided from the start since it is important that the roles and responsibilities of the client and contractors' representatives are clearly understood.

It is generally desirable for the client's operations staff to be involved in commissioning, even when this is being carried out by a contractor, and thought should be given to this when drawing up a contract.

When inviting firms to tender for a contract, the client organization must consider not only the contractor's competence and competitiveness in design and manufacture but also proposals for, and track record in, plant commissioning.

2.2 Types of contract

The importance of contract form needs to be appreciated whether a project is predominantly 'in-house' with perhaps relatively small involvement of subcontractors, or the entire work is to be entrusted to a contractor. Responsibility for commissioning needs to be clearly defined to avoid ambiguities at a later stage.

A wide range of types of contract exist covering all likely combinations from 'turnkey' to full reimbursable contracts. At the enquiry or bid stage of the project the definitions of responsibility must be carefully considered together with the general Conditions of Contract.

There are a number of model Conditions of Contract for process plants, including those published by the Institution of Chemical Engineers for lump sum, reimbursable contracts and subcontracts (see References 1, 2 and 3). They provide well-defined guidelines concerning related responsibilities, terminology and documentation appropriate to all stages of a contract, including commissioning. Whilst designed for work in the UK, they may also be used as a basis for preparation of Conditions of Contract for work in other countries.

Even if a recognized model form of Conditions of Contract is not used, it is nonetheless important that the contract defines the split of work and responsibility between client and contractor. This split of work will be significantly different depending upon whether the process technology is owned (or licensed) by the client or contractor. Table 2.1 contains extracts from a procedure developed by one UK process plant contractor to define such responsibilities. At the same time it identifies activities which interface between construction and commissioning. Procedures of this type may need to be adapted to suit the type and complexity of a particular plant.

Whatever form of Conditions of Contract is used, it is important that the requirements for completion of respective phases of the contract are clearly defined. Such milestones in the progress of site activities are usually formalized by appropriate certification. To these many may be linked to stage-payments, or transfer of ownership with its significant effect on insurance liabilities.

Terminology relating to phase completion at site needs to be specific, particularly in respect of precommissioning. Precommissioning usually commences during the latter stages of construction and has to be carefully integrated with them. 'Completion of erection' or 'mechanical completion' are two terms commonly used in model Conditions of Contract and frequently relate to the readiness of the plant (or part of it), to be commissioned using process feedstocks, or otherwise made 'live'. Thus all precommissioning is included by definition. In some instances this may not be so, stipulated mechanical tests having to be fulfilled before the completion of erection certificate may be submitted; precommissioning then follows.

2.3 Quality Assurance

Engineering of process plant projects increasingly involves the application of a disciplined management system to ensure that the quality of product or service is built-in at every stage. In this context, quality is defined as 'fitness for purpose' — that is, the product must be fit for the purpose for which it is designed.

It is important to establish at the outset of a project whether Quality Assurance is to be applied. Quality Assurance requirements are directly applicable to four key areas:

Table 2.1   Defining work and responsibility of construction and commissioning by contractor and the client*

Construction and precommissioning:	Phase A — Prepare plant/equipment for precommissioning/mechanical testing
	Phase B — Prepare services; clean and pressure test systems
	Phase C — Check and prepare major mechanical equipment, instrumentation, and protection systems
	Phase D — Final preparations for start-up
Commissioning:	Phase E — Charge with feedstock and so on. Start up plant and operate
	Phase F — Performance test and plant acceptance
	Phase G — Remainder of maintenance period

Phase	A	B	C	D	E	F	G
Control of works by:		Contractor construction				Client	
Basic responsibilities and phases for various categories of staff at site:							
Construction		Site manager and team of construction specialists			Defects correction under contractor's direction		Return for corrective work
Contractor comm		Witnessing inspection and testing			Advising start-up and operation of plant	Witnessing performance test	
			Training client's operators				
Client		Witnessing inspection and testing			Plant maintenance and normal adjustments by client		
			Operators receiving training		Preparing for start-up; starting up and operating plant		
Equipment vendors		Testing and commissioning presence on site			On call to site to end of maintenance period		
On completion of construction:							
		Certificate of mechanical completion/taking over certificate					
					Certificate of acceptance of plant performance		
							Final certificate

* See Appendix 2 for schedules listing principle work carried out in respective phases and by whom (pages 84–88).

- the design and specification of the plant;
- supply of equipment and services to the design specification by the contractor and all subcontractors;
- erection, installation and testing of the equipment to the design specification by the contractor and all subcontractors;
- operation of the completed plant to achieve design performance consistently.

Clearly, the first three areas relate to assuring the quality of the plant which is being built. Use at every stage of companies which have a proven Quality Assurance system and which may well be registered with an accredited scheme (such as that of the British Standards Institution BS EN ISO 9000) will ensure a high level of workmanship and minimize rectification work (see Reference 4).

Applying Quality Assurance to precommissioning commencing preferably with early involvement in the design phase, will prove beneficial and minimize problems that affect precommissioning. The same structured approach should then be applied to testing programmes at equipment vendors' factories, through to site construction, precommissioning and functional testing phases.

The fourth area listed above relates to how the plant is designed, controlled and operated to achieve consistent performance and the necessary corrective action to be taken when it fails to do this. At all stages of commissioning two questions need to be asked when changes are proposed:

- 'What effect, if any, does this action have on the ability of the plant to perform consistently within design specification?' (Any change must be referred back to the design authority and correctly logged);
- 'Does the change comply with the appropriate quality manual?'.

2.4 Commissioning estimates

A manpower cost estimate for commissioning staff is a usual element of the overall project costing for tender or budget purposes. It takes full account of all disciplines involved in commissioning, the hours including overtime that they may work, the back-up services that they need and the site allowances, travel and accommodation costs which they will be paid. Previous experience of similar plants is valuable for this purpose. Such information may be available as a computerized database.

It may be necessary to separate precommissioning costs from those for commissioning, particularly if planning is to be included in a firm or fixed price estimate and the commissioning is reimbursable. This sometimes occurs if the ultimate owner's process technology is involved.

It is important when putting together the budget for a project to make adequate allowance (either as a specific allocation or within contingency) to cover the cost of modifications which will inevitably be required during or as a result of commissioning tests.

When using a factored approach to estimating commissioning costs, plants or processes with a high content of 'new technology' attract a higher than normal factor. Similarly, a smaller project generally costs proportionately more to commission than a larger project.

Additionally, the costs of any temporary works should not be overlooked — particularly where time-related hire charges or service costs are involved.

2.4.1 Resourcing

The manpower estimate takes into account whether all resources are available in-house or whether it is expected to hire in supplementary personnel. When deciding upon the manning strategy for a project it may be necessary to consider all other company demands upon resources over the project timescale. Cost-effectiveness of available alternatives must be taken into account.

A further factor to be considered is the possibly disruptive effective of an imbalance between permanent and hired staff, although this is more directly relevant when actually selecting a commissioning team.

When a Quality Assurance system is included, provision for the greater involvement of commissioning personnel with all disciplines at a relatively early stage of the project needs to be taken into account. Such provision also includes preparation of detailed procedures and generation of associated documentation.

2.4.2 Programming

Project programmes should include detailed assessment of the duration and cost of performance testing, precommissioning and commissioning phases of the project. These must be aligned with definitions appropriate to the particular Conditions of Contract referred to earlier. Sadly, commissioning is too often given inadequate attention in the overall project estimates, with bad underestimating of both the time needed and cost involved. Similarly, it is too often the case that when slippage occurs in design or construction, commissioning time is arbitrarily reduced to hold the programme end date. Commissioning teams are accustomed to challenging targets, but the timescale and budget within which they are expected to complete their work must be at least realistic if they are expected to commit to it.

2.4.3 Personnel aspects

Costs relating to aspects such as inducements, status, travel, accommodation and living allowances must be considered, particularly in respect of foreign projects.

Inducements may need to take account of Conditions of Employment of the personnel likely to be involved, especially if field assignments such as commissioning constitute a departure from the norm and hence involve disturbance factors.

2.4.4 Crafts involvement

The construction element is not included in this guide except in relation to specific requirements for precommissioning and commissioning. The need for adequate numbers of supervisory and crafts personnel should not be overlooked.

Construction estimates usually take into account requirements up to completion of erection and these should include a small team of respective crafts disciplines with supervisors to work under the general direction of commissioning staff responsible for precommissioning. If such responsibility extends into the commissioning phase, it is prudent to make allowance for continuing involvement to cover remedial or rectification work as well as commissioning adjustments. Usually, a small team is required on shift (or at least available on call out) during start-up and initial operation of the plant.

Instrumentation usually requires the involvement of a significant number of technicians in all of these phases and on shift during start-up. It is sometimes necessary to differentiate between instrument fitters and instrument technicians. The particular skills of instrument technicians are in calibration, setting-up, loop-checking and optimizing settings on the installed equipment.

2.5 The project stage – early preparations

The first activities to involve commissioning personnel are likely to include:
- preparation of training programmes, where necessary;
- involvement in the design phase;
- check-listing of engineering and utilities line diagrams, (ELDs and ULDs);
- preparation of equipment check-out schedules;
- planning of precommissioning and commissioning activities for inclusion in the project and construction programmes.

2.5.1 Training

Responsibility for training 'key' members of the eventual operating and maintenance teams often rests with commissioning personnel under the direction of the project manager and in liaison with respective engineering disciplines (see Chapter 4). The extent of the contractor's responsibilities for training the client's operators and/or technicians must be clearly defined in the Conditions of Contract.

2.5.2 Involvement in design phase

Involvement in the design allows advantageous features to be included and problems that affect precommissioning, start-up and operation of the plant.

2.5.3 Preparation of check-lists

Commissioning staff may participate effectively in a review of line diagrams using check-lists prepared using information from experience of similar plants or equipment. This type of check corresponds in part to hazard and operability (Hazop) checks generally practised for new projects.

An effective retrieval system for experience gained in respect of similar plants or situations can be very helpful when reviewing procedures/designs for new projects and in resolving problems during commissioning.

2.5.4 Equipment check-out schedules

An audited Quality Assurance system should include a suitably structured approach to safe precommissioning and commissioning operations.

Check-sheets, or schedules, appropriate to the particular phase and activity should be available, commencing with those required during testing at the equipment vendor's factory, through to construction and precommissioning.

The check-sheets may be preprinted with standard questions annotated against process and engineering data sheets and vendor's drawings. They are signed-off by respective authorities after individual checks have been witnessed and eventually form part of the project handover documentation.

Examples of check sheets developed by one UK process plant contractor are included in Appendices 2.1 to 2.5 (pages 84–88).

Any procedures relating to specific activities of precommissioning and commissioning must be available with the check-sheets.

Statutory requirements will usually reinforce the need for proper completion of all verification documentation before a plant can be put into operation. The commissioning manager must resist pressures to take shortcuts in the supposed interest of expediting start-up.

2.5.5 Temporary works

The nature and extent of any temporary works should be reviewed during the early stages of project planning as these can have a major influence on the manner and approach to commissioning.

9

The following should be considered carefully:
- features which will aid commissioning — for example, additional branches to allow the hook-up of commissioning equipment such as pumps, instruments;
- temporary equipment — for example, additional pumps, by-pass lines, storage tanks, strainers and blanking plates;
- temporary supplies of consumables — for example, portable gas supplies, and feedstock (both primary and intermediates);
- temporary discharge routes (and the necessary authorizations for these) — for example, tankers and pipework.

2.5.6 Activity planning

The importance of careful, detailed planning of all phases of testing and commissioning cannot be over-emphasized.

Commissioning needs to be considered in good time within the overall project programme. Precommissioning activities have to be integrated into the construction programme to provide the best logical sequence for start-up preparations. Assessments must be made of when essential utilities and supplies (provided possibly by others) need to be available.

It is customary to prepare activity networks for precommissioning and commissioning programmed against target dates or durations. These may be in barline, arrow diagram or precedence diagram form, or sometimes a combination of all three. Figure 2.1 (pages 12–13) shows part of a typical planning bar chart.

A parallel planning requirement is to assess when commissioning staff should be assigned to site and for what duration. Construction progress needs to be regularly reviewed and the precommissioning programme updated. Assigning commissioning staff to site has to be matched with events which justify their presence — otherwise not only are the economics of the operation adversely affected, but impetus may be lost through inadequate work content. Conversely, unduly late arrival may handicap completion of erection as well as precommissioning. Contractors are rarely able to consider the requirements of any one project in isolation and regularly review overall commissioning commitments and staff deployment to meet what are often conflicting demands.

For overseas projects, matters such as the preparation of terms and conditions for staff also require early consideration.

2.6 Preparation and composition of commissioning team

If the project involves a new process, an induction course may be necessary for the commissioning team. Legislation, such as the UK Construction (Design and Management) Regulations and the Health and Safety at Work Act (and subsequent case law) emphasizes that operating staff should possess adequate knowledge of potentially hazardous situations including those which might be outside their previous experience. An induction course may consist of lectures by respective disciplines covering project, process, design and safety topics together with a visit to a working plant — preferably, this should take place not too far ahead of the planned start-up.

Whether or not a formal induction is arranged, it is customary for members of the commissioning team to attend a briefing before proceeding to site. For the commissioning manager this may involve an extended period to gain familiarity with the background of the project.

Ideally, the commissioning team should consist of engineers with relevant experience who have previously worked together and have sufficient depth of experience in all aspects of plant precommissioning and commissioning. A successful team may well include a balance between those of theoretical and practical ability.

Whilst technical competence has to be carefully considered when selecting the commissioning team, personality aspects are no less important. Choice of the commissioning manager must give consideration to the individual's abilities as a team leader; compatibility of team members also has to be taken into account bearing in mind the conditions of physical and psychological stress often experienced during plant start-up. Such demands may be particularly severe when the site of the project is in a foreign country.

The size of the commissioning team may vary with the type and complexity of the installation as well as with the contractual requirements and experience of the operating personnel. For example, the typical composition of a contractor's commissioning team on a large ammonia plant is:

- commissioning manager and deputy;
- four shift leaders;
- four assistants;
- instrument, mechanical and electrical specialists;
- chemist.

It is preferable to provide a four-shift rota to help avoid extended working hours. Critical start-up operations may necessitate a doubling-up of personnel to provide 12-hour shift cover for a period.

2.7 On-site activities

For the large ammonia plant referred to in the previous section, commissioning staff would normally be at site at least four months prior to start-up. A progressive build-up of the team would occur during this period. The stage at which continuous

Line	Task No.	Category Process	Activity Description	1	2	3	4	5	6
1	No.	Process							
2	12	"	Check out/line flush and pre. comm. equip. in crude unit area						
3	13	"	Check out/line flush and pre. comm. equip. in hydrotreat area						
4	14	"	Check out/line flush and pre. comm. equip. in reformer area						
5	15	"	Check out/line flush adn pre. comm. equip. in LPG unit area						
6	16	"	Flush fuel gas/fuel oil main						
7	17	"	Check out/line flush flare system						
8	18	"	Pressure test fuel gas/fuel oil main						
9	19	"	Pressure test flare system						
10	20	"	Load de-ethaniser column						
11	21	"	Load hydrotreater reactor						
12	28	"	Dry crude heater refractory						
13	29	"	Dry heaters – hydrotreater unit						
14	30	"	Chemical clean – compressor pipework						
15	31	"	Pre-commission and run in compressors						
16	32	"	Pressure test crude unit						
17	33	"	Pressure test hydrotreater						
18	34	"	Pressure test reformer unit						
19	35	"	Pressure test LPG unit						
20	39	"	Circulate crude oil in crude unit						
21	40	"	Heat crude furnace and commence tower circulation						
22	42	"	Commission desalter						
23	43	"	Commission stabilisor/fractionator on hydrotreater						
24	44	"	Excess fuel/LPG to flare						
25	45	"	N2 circulation on reformer						
26	46	"	Dry reformer reactors and refractory on fired heaters						
27	47	"	Charge catalyst to reformer						
28	48	"	Final pressure test reformer/purge with N2 and then H2						
29	49	"	Produce acceptable feed for reformer start up						
30	51	"	Circulate with H2 and heat reformer reactors						
31	52	"	Commission stabiliser on reformer unit						
32	53	"	Condition reformer catalyst						
33	54	"	Introduce feed to reformer						
34	55	"	Circulate N2 on hydrotreater reactor and heat						
35	56	"	Introduce hydrogen and feed to hydrotreater reactor						
36	57	"	Produce on spec. products from hydrotreater area						
37	58	"	Commission caustic and water wash vessels in LPG area						
38	59	"	Commission feed to LPG unit						
39	63	"	Trim units/conduct performance tests						
40									

Figure 2.1 Typical commissioning planning bar chart (part of)

Week No.																				
7	8	9	10	11	12	13	14	15	16	17	18	19	20	21	22	23	24	25	26	27

attendance of commissioning staff is required at site varies with the type and size of installation as well as contractual conditions. So does the rate of build-up of the team. The importance of thorough precommissioning of process plant in achieving a smooth start-up cannot be over-emphasized. Risks associated with shortcircuiting these activities to get the unit 'on-stream' include frustrations and sometimes hazards at the ensuing start-up.

Precommissioning activities under the direction of the commissioning team are carefully integrated with the completion of erection through close coordination with the site construction manager. Similarly, there is close liaison with the staff who are ultimately responsible for operation and maintenance of the plant.

Utilities are generally the first to be commissioned in order to provide facilities such as electric power, cooling water, demineralized water, instrument air, plant air and inert gas. Individual plant systems and equipment items are then checked for completeness against line diagrams, component drawings and schedules while lists are prepared of outstanding items (see Chapter 7).

During systematic checks commissioning personnel are also on the look out for features which may constitute a hazard to personnel or equipment — for example, fire risks from leaking flanges. The commissioning manager's brief is clear in respect of changes in design (see Chapter 3).

Following itemized checks all vessels are thoroughly cleaned and pipe lines flushed with air or water to ensure that debris is removed. When possible, water flushing operations follow immediately upon hydrostatic and mechanical strength testing of pipework. High pressure steam lines to turbines must be meticulously cleaned by repeated blowing through to atmosphere at high velocity against target plates to achieve the standard of cleanliness required before coupling to the driven equipment; usually this is carried out in the presence of the turbine vendor's commissioning engineer. Chemical cleaning of boiler systems, compressor connections and any special plant pretreatment is carried out as necessary.

Catalysts and packings are carefully loaded into vessels using established techniques and applying appropriate tests. Following these operations, air or nitrogen pressure testing up to normal working pressure is carried out to ensure system tightness. Specialist electrical, mechanical and instrument commissioning personnel concentrate on their specific tasks throughout this period. Machinery is run light and functional testing of instrumentation and control loops is carried out after calibration and setting up. Trip systems and other safety devices are carefully checked out in conjunction with process commissioning personnel.

Equipment vendor's commissioning personnel are called forward as appropriate, care being taken with regard to timing and providing adequate notice.

Throughout the precommissioning period it is highly desirable that operating and maintenance personnel of the production company or division participate actively in all that goes on. As well as providing an excellent opportunity to familiarize themselves in detail with the equipment procedures, it enables respective team members to get to know their opposite numbers before start-up.

2.8 Safety

Responsibility for safety should be considered carefully when drawing up contracts and should be defined in the Conditions of Contract. In drawing up contracts for foreign projects, local laws relating to safety and liability for equipment and personnel must be taken into account.

Safety is considered in detail in Chapter 3; aspects of particular relevance are:

• ensuring that safety considerations are paramount in plant design and operation. The UK Construction (Design and Management) Regulations require the preparation of a detailed health and safety plan, including commissioning, before the start of construction. The UK Health and Safety at Work Act highlights the need for safety audits in plant design and regard for safety is the responsibility of the individual at all stages of design and operation;

• safety awareness must not be sacrificed for objectives which may appear to assume greater importance as the commissioning stage approaches. Regular safety checks should be made and attention given to the special precautions required if construction work is to continue in the vicinity of the plant to be commissioned. Escape routes from potentially dangerous areas must be unimpeded and manual trip devices easily accessible in the event of an emergency. Suitable alarmed detectors and escape masks/survival equipment should be available as appropriate when leakage of toxic or other harmful gases and so on may be encountered.

2.9 Site liaison and communication

Misunderstandings and possibly a breakdown in communications can arise between the commissioning team and operating staff unless sufficient attention is given to liaison. This is particularly important on foreign projects where day-to-day activities are generally on a more formal basis and often handicapped by language difficulties. Liaison methods should cater for conveying technical advice and response, via interpreters if necessary, and also provision of joint logging of plant operations. When working through interpreters it is good practice, particularly in respect of critical operations, to require that the message be repeated back to the originator before translation.

It is also important to establish lines of communication between day and shift operations. Where a commissioning team is responsible for directing operations, it is customary for directions to be routed through nominated technical personnel of the operating company who in turn supervise their own operators. Even under these circumstances, however, the commissioning team may need to become directly involved in plant/process adjustments. This is most frequent during start-up and emergency situations.

It is important that formal communication between client and contractor is through a single clearly identified channel, except in the case of safety when any unsafe activity should be stopped immediately.

Adequate means of communication between the control room and personnel in field locations must be tried and tested in good time before commissioning begins. Liaison with local authorities and the local community should be maintained in order to ensure good relations.

2.10 Performance testing

Provision for plant performance testing requires careful liaison between the appropriate personnel of both parties. Broad requirements of such tests will usually have been stipulated in contract documents but it remains for site teams to schedule the test arrangements in detail. These are likely to include the selection and calibration of measuring devices together with the application of correction factors and tolerances, sampling and analyses, datalogging and effects of interruptions. It is important that procedures are drawn up and agreed upon well in advance of proposed test periods.

2.11 Operational training

On-plant training of operating personnel generally continues throughout start-up and performance testing. One area which usually features prominently at this time is actions to be taken on the activation of safety trip systems. These either pre-alarm impending shutdown of one or more sections of plant requiring prompt corrective action, or may initiate an immediate trip-out. Operating staff should become proficient in dealing with these and other emergency situations. Proficiency requires adequate training, not only during initial instruction but by regular refresher sessions after the commissioning team has completed its involvement.

Safety, health and environment

3

The commissioning period is often the most difficult and potentially hazardous phase in the life of a process plant. Both client and contractor have a clear legal obligation to ensure the safety and wellbeing of their workforce, the public and the environment. The best way to discharge this obligation is by attention to detail, planning all commissioning activities with care, assessing all potential hazards in depth, providing protective and remedial equipment and, above all, investing in appropriate training for personnel. This ensures that they are all 'suitably qualified and experienced' for the duties that they have to carry out.

3.1 People

To achieve safe, effective commissioning, it is essential to use a well-trained, technically strong team with relevant experience. The technical staff requirement is roughly twice that for normal operation. Sometimes it has to be accepted that in order to provide these resources other plants/departments have to release good people for a time to get through this critical period. Experienced contractors can often help either by carrying out the whole operation or by supplementing the client's personnel. Companies have a legal obligation to demonstrate that competent staff are employed.

3.2 Safety training

3.2.1 Commissioning team

The training of the commissioning team (see also Chapter 4) in safety matters is most important. All members of the team should be aware, to an appropriate degree, of the plant design philosophy and the potential hazards, both process and engineering. Where possible, key commissioning personnel should be involved from the early stages of the project. Part of their remit is to consider the safety aspects of the plant, especially the start-up phase and emergency shutdown phases. This knowledge can then be passed to the rest of the team when they assemble.

3.2.2 Operating team

The safety training must be comprehensive and properly integrated into the total training programme. All personnel should already have had some safety training as part of their induction to process operations, including Permit to Work, Vessel Entry and other formal procedures, with special training on the potential hazards given as appropriate. It can be useful to involve all grades of personnel in producing the emergency procedures including power failure, steam failure and so on. Shift teams are encouraged to discuss how they would deal with emergencies which might occur when there are no day management staff to take charge. Practical experience at first hand of operating similar processes or equipment is most useful. A model (physical or computer generated) of the plant is also very useful — for example, for checking means of escape, locating safety equipment and checking safe access. The commissioning team reviews any safety reports (Hazop and so on) that have been carried out and is made aware of the hazards and types of materials being processed and handled by the plant, including all effluent streams.

Special discussions also take place where appropriate with the factory central safety organization, the Health and Safety Executive, and with the local authorities and the Environment Agency on how to deal with emergencies such as fire, toxic hazards and so on. Joint exercises can be very valuable to try out agreed actions.

In addition to all this, plant personnel need regular training in fire drill. It is good practice to issue a pocket card giving the basic actions in the case of fire and other emergencies. Fire and first aid teams should be trained in any special procedures relevant to the new plant or process and should become familiar with the treatment procedures to be followed after exposure to the chemicals used in the process.

The precommissioning period is the ideal time for this training, especially when the team has access to the process plant and can get a feel for the equipment, layout and so on. Familiarization helps enormously in hazard avoidance.

3.3 Planning

Planning for safe commissioning is essential; a well-planned start-up tends to be a safe one. The degree to which the planning is carried out is dependent on the size of plant or nature of the process. In the past, safety matters were often given a low priority and consequently forgotten or omitted because of more pressing or interesting technical matters. It is vital to ensure that 'Safety Really Does Come First'. A plan should be made of what has to be done when and who will be responsible. It must be understood that safety preparation, if not carried out, can hold up commissioning.

3.4 Communications

Effective communication is essential for minimizing hazards in process plants. For example:

- written — the operating, safety, commissioning and other instructions should be carefully written and supplemented by clear diagrams to make sure they are easily understood. They must cover variations in process operating conditions;
- audio — communications between all grades of operating personnel on the plant are most important. Consideration must be given to: two-way radios, loud hailers, pocket pagers ('bleepers') and internal telephones. Choice depends on local conditions. Effectiveness is more important than cost;
- visual — appropriate safety notices and warning signs must be considered (many will be mandatory);
- audiovisual — aids such as films, video tape and tape/slide sequence methods should be considered for communicating important safety messages. Some might be general — for example, dealing with fires — and some might be specific to some part of the process.

Finally, it is hard to beat good old-fashioned dialogue between people, at all levels.

3.5 Plant handover

Usually the construction people are anxious to get the plant formally handed over for commissioning. It is important that short cuts are not taken in haste to complete handover.

The primary objective of both the construction and commissioning teams is to get the plant safely into earliest beneficial operation. It therefore helps to have mutual cooperation so that some of the commissioning team have access to the plant as it is being built. They are able to see the plant grow and therefore have an intimate knowledge which helps towards safe operation later. It is also possible to get many items — such as direction of atmospheric vents, poor access to valves and instruments — cheaply rectified before construction is complete. After that changes are expensive and time-consuming (see Chapter 3,

Section 8, page 22 on modifications). The construction team should give reasonable warning of their intention to offer the plant for mechanical acceptance and preliminary reservation lists, which include safety items, must be drawn up. A typical check-list is shown in Appendix 3.1 (page 89). In this way, many items can be cleared up informally. At handover, the plant should be checked line by line, valve by valve, against the 'Approved for Construction' process and instrumentation diagrams (P&ID) to ensure that the plant meets the design intent. The formal handover usually has a certificate with the final reservation list of remedial work required. It is most important that the plant is handed over in a safe condition, correctly built and to design.

With the advent of big plants and the frequent need to get plants of any size quickly on-line, a method of selective handover of plant systems has often been adopted. These are usually utilities, such as cooling water, steam mains, steam raising boilers and so on. Preferably these are treated in separate geographic areas, but this is not always possible. The order of handover is determined by mutual agreement to cover those process areas which are required earlier, or those where because of deliveries early completion is easily achieved. Naturally, commissioning of a system surrounded by normal construction work is not lightly tackled because of the potential safety hazards involved. Construction personnel accustomed to having relative freedom of working are not usually aware of the hazards or the procedures involved in plant commissioning and so clear procedures and education are needed to ensure safety.

Careful planning and communications are required at all levels. A clear definition of what is handed over from construction has to be made — for example, with formal marking up of line diagrams, segregation barriers complete with clear signs, clear labelling of pipe every few feet and insertion of clearly labelled slip-plates. Once equipment is formally accepted by the commissioning team, all work must be carried out using the full safety procedures of the process plant operators — for example, using Permit to Work Certificates and Entry Certificates.

3.6 Getting the plant ready for commissioning

Usually the control room building (not necessarily complete with instrumentation) is one of the earliest areas to be handed over and it is from this centre that the plant preparation and commissioning work are organized. One of the first tasks before any work starts is to check that all the safety equipment is installed, and the emergency and safety procedures are understood by all concerned.

The next main task consists of cleaning and proving by putting a flow of water or air incrementally through every pipe, valve, fitting and so on at a rate as near as possible to that which will be sustained in actual operation. Where there are non-commissioning team people in the area, these operations often have to be carried out at 'safe' times — for example, at meal breaks or after day hours. All this preparation progress has to be carefully marked up on master line diagrams to check that nothing is missed. Very special attention is paid to ensuring that the lines on each side of relief valves and bursting discs are clear.

After washing or blowing the systems, and carrying out any necessary drying, the plant is usually leak tested by pressurizing with compressed air. This is a most important stage from the hazard point of view since the objective is to ensure there are no 'holes' in the system which could lead to leakage of flammable, toxic, corrosive or otherwise noxious material.

At this stage a slip-plate ('spades' or 'blinds') register is compiled which states where slip-plates are located in the process. The register must, of course, be kept up-to-date, and signed by the checker.

Very often air has to be purged from a system with an inert gas — for example, nitrogen. There should be very strict rules for the admission of inert gas into a plant. Basically, none should be allowed in without the express authority of the commissioning manager, who has to be satisfied that all the process equipment has been handed over and therefore entry rules apply to all vessels.

Another important issue in the progress of commissioning many plants is at which point flammable materials can be admitted to the system. Before this can be authorized, the commissioning manager must be satisfied that there are no unauthorized sources of ignition. This means, for example, that all burning and welding work must be complete, no blow lamps or tar boilers can be used, and smoking is prohibited. Under very special circumstances, such equipment can be employed in conjunction with a Fire Permit, which is issued after certain very special conditions and precautions are undertaken. Appropriate precautions must also be taken for other noxious materials.

3.7 Initial commissioning

Before the final commissioning of the plant, which is usually the introduction of feed, it is useful to try out as much of the plant as possible with as little risk as possible. For example, some pumps can be run with water if this is mechanically satisfactory. Discrete sections of the plant might be tried with material as near as possible to subsequent operating conditions, shutting them down again, if necessary, when proven. Malfunctions usually develop in the first few hours of operation and it is safer to identify and handle them with concentrated effort, avoiding the temptation to start up too many items of plant at once. Special equipment — for example, compressors and steam boilers — should be started up and shut down by all shift teams to gain experience.

Before process commissioning, the commissioning manager should lay down two more safety rules:

- the plant must be cleaned up to the commissioning manager's satisfaction, which usually means that all rubbish, trip hazards, non-essential scaffolding and other materials (especially flammable or toxic) should be cleared away;
- the person in charge of commissioning a system has personally checked the plant condition to ensure all is safe and correct. Appendix 3.2 (page 90) shows an example of a safety check-list before commissioning a plant is given.

It is good practice to carry out a 'prestart-up' safety review, incorporating a comprehensive plant walkround and a review of the minutes of all previous safety reviews.

3.8 Modifications

During commissioning, the need for plant modifications will arise. These modifications are potentially the greatest hazard to a new plant. Experience, including major incidents, teaches that all modifications must be subject to a strict approval system to ensure that design and safety standards are preserved.

This must include a Hazop (or equivalent) study to determine the implications for the rest of the plant. For example, the insertion of a valve may create a blocking in or an over-pressure problem. The authorization must be given in writing by a competent person. For larger plants, a commissioning modifications engineer can be appointed who stands apart from the normal pressures of the commissioning team. An example of a check to be used in the safety assessment of a modification is given in Appendix 3.3 (pages 91–92).

3.9 Quality Assurance

Where a Quality Assurance system (see Chapter 2) is employed on the project, much of the foregoing should be considered as part of the discipline.

3.10 Legislation

As with all other industrial and commercial enterprises, the design, construction, commissioning and operation of process plant are subject to national and in the UK, increasingly to European legislation.

The principal features of the UK safety, health and environmental legislation relevant to process plant commissioning are outlined. It must be emphasized that this guide can neither include comprehensive details of all UK regulations nor the equivalent legislation of other countries. As a priority during the early planning of commissioning, those responsible are strongly advised to study safety, health and environmental legislation carefully to ensure that the planned commissioning activities comply and that all necessary authorizations are sought and obtained prior to commissioning.

Perhaps the corner stone of Health and Safety legislation in the UK is the Health and Safety at Work Act 1974. This Act lays an unequivocal obligation on employers: 'It shall be the duty of every employer to ensure, so far as is reasonably practicable, the health, safety and welfare at work of all his employees'.

3.10.1 Safety legislation

Before the start of commissioning, routine safety testing and certification is completed — for example, for pressure vessels and cranes and other lifting equipment. Commissioning is the time when non-routine operations are likely to be necessary. It is imperative that personnel are properly trained and regulations are followed when, for example, confined space entries are required and that adequate procedures to ensure the safety of personnel, are in place and followed when mechanical or electrical equipment has to be operated with normal guards or covers removed.

The Construction (Design and Management) Regulations

In response to the generally poor safety record of the construction industry, all-embracing legislation (The Construction Design and Management (CDM) Regulations 1994) has been introduced. The CDM regulations have changed the way in which commissioning activities are approached; commissioning is defined as a construction activity under CDM. This means that designers now consider the hazards faced by commissioning teams and, wherever practicable, deal with these hazards by improving/changing the design of the plant — for example, by specifying see-through guards, 'dead-man' controls and 'inch' buttons.

The client is obliged to ensure the competence of contractors (specifically the principal contractor) and designers and that adequate resources are allocated to health and safety. A planning supervisor is appointed to co-ordinate the health and safety aspects of the design. This role is a key one and it is vital that the planning supervisor and the designers have a sound knowledge of (a) all aspects of construction safety and (b) the rules which govern safety during commissioning (for example, the rules concerning working at heights and governing working on live electrical equipment). The client provides the planning supervisor with all information relevant to health and safety on the project and ensures that construction work does not start until the contractor has prepared a satisfactory health and safety plan. Commissioning activities must be incorporated into the health and safety plan from a very early stage in the design, particularly if it is anticipated that construction activities are to be carried out alongside commissioning work.

It is vital that commissioning teams are fully familiar with all aspects of construction safety; all necessary training programmes, site experience and so on must be completed before commissioning personnel are allowed to commence work on site (formal training includes training on hazardous substances, noise, moving machinery and working at heights).

Commissioning team leaders should study the CDM regulations and liaise closely with the planning supervisor to ensure comprehensive planning of the safety-related aspects of commissioning.

The Control of Industrial Major Accident Hazard (CIMAH) Regulations

If the process inventory exceeds certain defined limits of flammable or toxic materials as specified in The Control of Industrial Major Accident Hazard Regulations 1984 (CIMAH), then before the process can be operated or commissioned, a safety case must be prepared and submitted (at least three months before the planned start of commissioning) to the Health and Safety Executive and accepted by them before commissioning can commence.

Following the adoption by the Environment Council of Ministers in December 1996 of the 'Seveso II' Directive, the CIMAH regulations will be replaced. A consultative document is expected early in 1998 with the final regulations coming into force in February 1999.

3.10.2 Health legislation

It is impossible in a guide to process plant commissioning to cover any of the specific legislation such as the Carcinogenic Substances Regulations, enacted to protect the health of workers. Nor is it possible to give guidance about the toxicity of specific hazardous materials.

There is, however, one overarching piece of legislation (The Control of Substances Hazardous to Health (COSHH) Regulations 1994) which sets out clearly the obligations of an employer to protect employees from the effects of hazardous materials.

The COSHH regulations require employers to:
• identify all hazardous materials to be handled and fully assess the risks arising from all planned activities;
• in light of the assessed risks, determine the precautions needed to avoid these risks;

- put in place and maintain the appropriate control measures to prevent or control exposure to hazardous materials;
- monitor the exposure of workers to hazardous substances and carry out appropriate health surveillance;
- ensure that employees are all properly informed, trained and supervised.

3.10.3 Environmental legislation

The introduction of the Environmental Protection Act 1990 (EPA '90) has focused attention on the environmental impacts of all aspects of life. This is clearly significant for process industries which are asked to consider the impacts of all environmental discharges in an integrated manner. By considering aerial discharges, liquid effluents, and solid wastes it is possible to achieve the optimum environmental solution.

These concepts of integrated pollution control (IPC) and best practicable environmental option (BPEO) are fundamental to the Prescribed Processes and Substances Regulations 1991 and subsequent amendments. The regulations define both the processes to be controlled, and the appropriate regulatory authority. Since the introduction of the Environment Act 1995 in the UK, the regulatory authorities are the Environment Agency (EA) in England and Wales, and the Scottish Environmental Protection Agency (SEPA). For processes with limited aerial discharges (Part B Processes as defined in the regulations), the local authority air pollution control unit is the appropriate regulator.

Discharges to sewers and inland or coastal waters are controlled by the Water Industries Act 1991 and the Water Resources Act 1991, respectively. Authorizations under the Water Industries Act are obtained from the local water company (for example, Yorkshire Water). The Water Resources Act is now regulated by the EA and SEPA, as it naturally falls into the integrated view of environmental process control.

All discharge consents must be obtained before commissioning can take place. Ideally, the regulatory authorities will have been involved in the project from an early stage. Applications for authorization must include justification of BPEO and demonstrate that the best available technology not entailing excessive cost (BATNEEC) has been applied. If these concepts have been put into practice during the design and development of the process, the authorization acquisition process should be relatively uncomplicated.

When making applications for discharge authorizations the commissioning phase should be specifically addressed. It is not uncommon for start-up conditions to affect the quality and quantity of process effluents dramatically and therefore it is possible that conditions of an authorization derived from expected steady-state flowsheet conditions may be too restrictive to enable the process to be commissioned.

The integrated view of environmental control is now common throughout the European Community since the introduction of the Integrated Pollution Prevention and Control (IPPC) Directive. As this is a Directive, however, it means that national laws must be developed to implement its requirements. Therefore it is possible that in some countries regulation may differ slightly, although working on the same principles. Similar regulatory systems are being established worldwide, and the concepts of BPEO and BATNEEC should be fundamental to them all. It is always important to understand fully as early as possible the specific national regulatory requirements that apply, and the process for obtaining the necessary authorization.

3.11 Postscript

One word summarizes the general philosophy of minimizing hazards during process plant commissioning. That word is check and, if in doubt, re-check and seek advice.

Training

4

It is important that all personnel involved in plant commissioning are carefully selected and given planned appropriate training which generally involves a combination of classroom, theory and 'on the job', practical training. This requires the transfer or recruitment of people in sufficient time for training to take place. Where possible the organization chart should be set out early so that the teams as they form can learn to work together. Ideally, some senior technical operating and maintenance staff representatives, and especially the commissioning manager, will have been involved from the initial design stage. They should be experienced people who are then in a good position to oversee the training of the rest of the team. It is increasingly common to use some form of psychometric or team-role analysis to help in team selection and building, and to undertake formal team building exercises off-site.

The preparation of troubleshooting guides provides a very effective training exercise for commissioning supervisors and can be of great value later, such as when problems develop in the early hours during a night shift.

Training of the total team is usually best organized in appropriate groups — for example, process operations, maintenance (mechanical, electrical, instruments) and laboratory.

4.1 Process operations

Normally the most intensive training is given to the process operating team. The managers and supervisors are appointed and trained first. Once trained they can in turn participate in the training of the process operators. Supervisors can also be usefully employed in drafting/commenting on operating, emergency, plant preparation and other instructions. Not only does this give effective learning about the plant but it also increases the sense of purpose and commitment. Training programmes for supervisors and operators should contain a large amount of self-learning. For example, they can draw up their own simplified process diagrams, prepare and give lectures to the rest of the group, and take part in informal discussions. The formal part of the training can typically include:

- induction and safety — a general introduction to the plant, including the basic safety and site emergency procedures (see also Chapter 3);
- plant layout — a study of the area in which the particular team will work, locating and identifying all main items of equipment. A plant model is very useful here;
- detailed operation — an in-depth study of the operation of the process systems, including familiarization with appropriate process and instrumentation diagrams, operating instructions and so on;
- emergency procedures — failure of power and other key utilities, vital equipment failure, hazardous leaks and other emergencies are considered. Actions to be taken should be discussed, understood and agreed;
- isolation for maintenance — methods for the safe preparation of equipment for maintenance, particular hazards and difficulties are emphasized.

4.2 Maintenance

In general, less formal training is required for the maintenance as opposed to the process discipline. All personnel should be given the basic safety and emergency procedure instructions as well as an appropriate appreciation of plant operation. More specialist training on new equipment can sometimes be given at the vendor's works, or by special local arrangements offsite. This is particularly important with instrumentation.

It is increasingly common for suppliers to provide fully detailed instruction manuals, particularly for maintenance of their equipment. This requirement must be detailed in the invitation to tender if claims are to be avoided.

Tradesmen require basic safety, emergency and process appreciation training. They should also be given special instructions about the equipment they will have to maintain and especially 'troubleshooting' potential problems.

4.3 Laboratory and specialists

After the basic safety/process appreciation training, personnel are given training on analytical equipment including sampling methods (especially where new technology is involved). This training includes practice to assure repeatability and accuracy in analysis.

4.4 Training methods

Once the training needs have been established, the methods can be selected from:

● custom-built courses, usually in-house and best carried out by the plant technical staff and engineers. Sometimes the company training department will be involved;

● courses arranged by outside organizations — for example, instrument vendors;

● training on other similar operating plants;

● use of training aids — for example, process simulators, (often computer controlled), films and videos. Some can be bought as packages, or produced inhouse. These can be particularly useful since they can often be available on a 24 hours a day for 'self teaching'.

4.5 Timing

All too often, personnel are brought in too late and are inadequately trained before start-up. Senior management should give adequate consideration to proper training of personnel for a safe and efficient start-up. Much will depend on the size and nature of the plant.

Technical staff need to be available as soon as they can be usefully employed, such as in assisting with the plant design (including overseeing a contractor), and participating in hazard and operability studies.

Supervisors should be brought in early enough for them to be trained in time to assist in the subsequent training of their teams. For large plant, inexperienced operators will probably require at least six months' training, including experience on other plants. The training time for tradesmen can be as little as one month where no new technology is involved, but far longer for specialized equipment.

Mechanical completion and precommissioning

5

Mechanical completion is the term used to cover the phase between equipment installation and the start of process commissioning, in which components of plant are proved to be mechanically fit for their process duty. It can be regarded as a specialized part of the precommissioning activity in which each component is prepared for process commissioning under the control of the commissioning team. A main contractor or equipment supplier is usually responsible for this phase of the work.

In practice the stages of commissioning are rarely so well defined. From the contractual point of view, mechanical completion occurs between 'completion of construction' and 'acceptance'. It may include some specified performance tests and usually refers to individual components of a plant rather than the total plant. This is dealt with in Chapter 2 and Appendices 2.1–2.5 (pages 84–88).

Since installation may be continuing in some areas of plant while others are being tested and commissioned, site safety is given detailed consideration; for example, component suppliers and subcontractors are carefully controlled during this phase since areas can change classification during the course of construction and commissioning.

Generally, precommissioning refers to preparing the plant for the introduction of process materials and its main object is to eliminate any problems which might arise at later and more critical stages of plant operation.

5.1 Mechanical completion

The sequence of mechanical completion is governed by the overall programme but usually starts with electrical power and utilities.

5.1.1 Objectives

The objective of mechanical completion is to prove that an installed plant component is suitable for commissioning. This phase includes:

- checking that equipment is installed correctly; this is usually carried out against check-lists that are produced from the component schedules and flow-sheets. For guidance, a typical example is given in Appendix 5.1 (page 93);

- proving that the basic components of equipment operate mechanically as specified, or at least acceptably for commissioning;

- demonstrating that instruments and control equipment work;

- proving to the commissioning team that components are suitable for precommissioning.

5.1.2 Administration

Paperwork is both a tool and a torment in all commissioning. During mechanical completion specialized paperwork relating to units of equipment or components is needed, for technical, safety, contractual and legal purposes.

Specialized requirements are:

- Inspection — this should be carried out before testing is allowed. It must be planned against check-lists and may require specialists (such as weld inspection). As a guide, typical check-lists are given in Appendix 5.2.1 and 5.2.2 (pages 94–96).

Checks are made against the specification and vendors' and contractors' drawings both for equipment details and system completeness. Systematic recording of checks is particularly useful where several similar units are supplied. Working drawings are key drawings for system checking and subsequent marking of a 'master' set of drawings can simplify recording.

Process safety and operability are checked — for example, orientation of relief valves, 'fail safe' systems, orientation of non-return valves, access to valves and adequate provision of drains and vents. The plant is finally checked for cleanliness, removal of construction debris and so on.

Physical safety aspects, completeness of safety fittings, handrails, means of access, escape routes, emergency showers and eyebaths, fire extinguishers and so on are also checked.

- Testing of components by construction staff — all tests of the component parts should be witnessed. The object of this stage should be to obtain any completion of erection certificates as set out in the Conditions of Contract, and to ensure that mechanical completion is feasible. Note that in some cases the warranty period runs from this date, in others from the date of commissioning.

- Paperwork for tests on each plant item, on pipework and instrument systems — test sheets should be prepared in advance, listing all the tests required together with space for entry and certification of results. Some examples of these proformas are given for guidance in Appendices 5.3.1 to 5.3.4 (pages 97–101).

• Paperwork required for handover from the construction team to the commissioning team — to speed up commissioning it is advantageous for the commissioning team to accept the plant from the construction team in sections, so that plant testing and checking can proceed as soon as possible. The commissioning team accepts full responsibility for any section of the plant after handover. This is when all the documents and certificates of tests already carried out by the construction team or by the original equipment manufacturers must be available.

During this handover period the commissioning team personnel record their reservations on plant acceptability for correction by the construction team. It is essential that during this period standardized paperwork is used, otherwise misunderstandings can occur. It is important to define the responsibility for accepting satisfactory completion of items on the reservation list.

5.1.3 General considerations

• Checks — before pressure testing, a functional inspection of each system is carried out to ensure that it has been installed correctly — for example, correct valve types, control valves the right way round, correct instruments fitted and so on. All vulnerable equipment such as pumps and control valves are fitted with temporary strainers. The system is then blanked off at each end, filled with water and pumped up to test pressure. The line/vessel test pressures must be compatible. Relief valve set pressures are tested on-site and tagged before installation. If necessary, relief valves are blanked during pressure testing of lines.

Sometimes — for example, where water could adversely affect the process or the materials of construction — a pneumatic pressure test is carried out. Pneumatic testing requires special safety precautions because of the hazard from stored energy should any part of the system fail. If a pneumatic test is unavoidable, specialist safety advice should be sought.

After the mechanical integrity of each system has been proved, it is flushed through very thoroughly and drained out. The instrument air supply manifold is normally pneumatically tested. It is essential to blow the air supply pipework through very thoroughly after pressure testing to remove all pipe debris.

• Instrumentation (see Chapter 6) — during commissioning, problems are often encountered with the instrumentation. Every effort must be made to ensure that the instruments are functioning accurately and correctly before start-up and have not been damaged during commissioning.

• Electrical — all earth continuity is checked and an acceptable level of grounding resistance of the structure and vessels achieved. Wiring is checked for continuity and insulation resistance. Finally, motors must be checked for correct direction of rotation. As a guide, examples of check-lists are given in Appendix 5.4.1 and 5.4.2 (pages 102–103).

● Water trials — before running the plant with process materials, it may be necessary to test it as far as possible by running with water. This also provides the opportunity to give the plant a final thorough flush through. Complete process simulation may not be possible. Water trials allow the majority of systems to be checked out, however, and they may allow the plant to be run hot, which further identifies leaking joints. It is essential to make use of the water trials to check the calibration of all the flow measuring instruments.

5.2 Precommissioning

The objective of the precommissioning phase is to prepare the plant for process commissioning proper. It may involve individual unit operations or a process area containing several pieces of plant.

Apart from the guidance given in Appendix 5.1 (page 93), the following specific points should be carried out prior to plant commissioning:

● check thoroughly the plant as built against the relevant plant flowsheets;

● ensure that the correct maintenance manuals are available;

● check that the Standard Operating Procedures (SOPs) are available and up-to - date;

● carry out simulated runs to ensure that all items of equipment function properly and that the control and instrumentation systems operate correctly (see Chapter 6);

● check that the necessary laboratory facilities are adequate and that the laboratory staff are trained to carry out the required quality checks on raw materials, products, and byproducts;

● ensure that all necessary chemicals and raw materials are available on-site and are of the correct quality and quantity;

● check that plant operators have been briefed and that they understand the operations of the plant systems.

Control systems

6

Whilst some older plants do not have computer control systems, most new plants and retrofits do. The emphasis in this chapter, therefore, is on the commissioning of control systems that are computer based. Most of the principles and procedures outlined, however, are just as applicable to analogue systems as to digital ones.

No attempt has been made to distinguish between different types of system: distributed or integrated, PLC-based or otherwise. The guidance in this chapter is essentially generic.

6.1 Timescale

The sequential/parallel nature of the various testing and commissioning activities is summarized in Table 6.1 and shown in Figure 6.1 (page 36).

Note that the testing of the computer control system and the precommissioning of the field instrumentation are independent activities. It is not feasible, however, to commission them separately because of their functional interdependence.

Much of the instrumentation cannot be installed until after the plant itself has been largely installed, and the computer control system is often one of the last items to be delivered to site. Consequently there is an overlap between the installation of the control system and the commissioning of the rest of the plant. This overlap needs careful management to ensure that the integrity of the control system is not prejudiced in the supposed interest of the overall commissioning process.

6.2 System testing and installation

Many computer control systems are purchased as 'turnkey', in which case most of the testing of both the hardware and software is carried out in the supplier's works, where the necessary expertise and facilities are readily available. Such

Table 6.1

Activity	Aspects
Field wiring/instrumentation	• analogue controller • sensors, transducers, transmitters, actuators • switches/relays/solenoids/contactors
Computer hardware	• power supply/watchdog • card/rack/frame assemblies • operator stations, peripherals
Systems software	• communications links • operating system • utilities software/tools • systems diagnostics
Input/output (i/o)	• i/o hardware • analogue/discrete/pulse • i/o database/engineering units
Configurable software	• database • signal conditioning blocks • alarms/trips/interlocks • analogue/discrete faceplates • overview/trend displays and so on • alarm lists
Applications software	• sequences • applications diagnostics • mimic diagrams • management information programs

testing is against agreed acceptance criteria and normally results in payment of most of the system value, the balance being paid on completion of site commissioning.

The testing of applications software by the supplier is usually effected by means of test boxes consisting of switches, lamps, voltmeters and so on. In recent years it has become more common for applications software to be tested by simulation techniques, an approach which can lead to significant savings in time due to the scope for increased parallel activity. This is discussed fully in Reference 5.

It is important to appreciate that such acceptance testing of turnkey systems is only part, albeit a major part, of the commissioning process which embraces functional checks on the whole system, including field instrumentation, plant interfaces, operator interfaces and support systems. This chapter concentrates on the post-acceptance, site-based phase of commissioning.

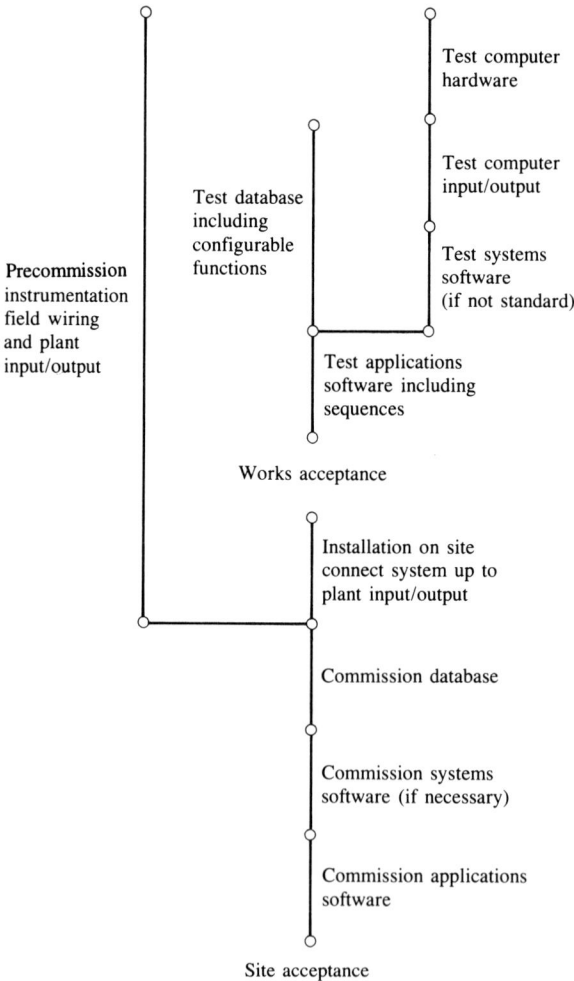

Figure 6.1 Sequences of activities in commissioning a computer control system

Site installation of the computer control system is, in principle, relatively straightforward and is usually carried out by the system supplier. It essentially consists of:

- installing the system cabinets and input/output (i/o) racks in interface rooms;
- connecting up the i/o racks to prewired termination cabinets, some of which may be in a separate motor control centre, usually by means of multichannel cables and connectors;
- positioning the operator stations and peripherals in the control room and connecting up their serial interfaces;
- powering up the system and carrying out standard system diagnostics routines.

6.3 Calibration

Instrumentation is always calibrated by the manufacturer but is often recalibrated on arrival at site, especially if there has been obvious damage in transit/storage. When recalibrating it is important to remember to:

- calibrate over the whole range and set the bias correctly;
- use the correct static conditions when testing differential devices — for example, differential pressure (dp) cells;
- use the correct process medium when calibrating analysers;
- check the local/remote switches function properly;
- ensure that any contacts for initiating alarms are set correctly.

Clearly, it is necessary to have access to the various test rigs and equipment required for calibration purposes.

6.4 Precommissioning of field instrumentation

Ideally the field instrumentation and process connections — for example, impulse lines for dp cells — should have been properly installed prior to commissioning as should the infrastructure such as power and air supplies, conduits/trunking/cable trays, wiring of termination cabinets and tags. In practice this is rarely so, and the primary objective of the pre-commissioning process is to identify faulty installation. If quality assured procedures have been used by the installation contractor, credit may be taken for this and some of the precommissioning checks omitted.

For detailed guidance on most relevant aspects of process, pneumatic and electrical installation practice, the reader is referred to BS 6739, Reference 6.

The emphasis in precommissioning is on checking. An essential first step in the precommissioning of instrumentation is a visual inspection of the installation such as its mounting, associated pipework and wiring, and to check that the workmanship is of an acceptable standard.

The principal check that always needs to be made for each element, as installed, is that its calibration, with regard to both range and bias, is consistent with its specification and duty.

Whilst, in general, the major contractor has overall responsibility for the precommissioning of instrumentation, there is much to be said, for familiarization purposes, for involving the site instrument maintenance personnel in the precommissioning process.

Subsequent checks, to be carried out systematically throughout the system, include ensuring that:

• the primary sensors installed are of the correct type and consistent with their transmitter ranges — for example, orifice plates and dp cells;

• the process interface is sound — for example, no leaks — and the correct orientation and right sealing fluids are used on impulse lines;

• the correct pneumatic/electrical interconnections have been made;

• the grade/quality of all tubing and cabling, single core or multicore, is appropriate to the duty, such as proximity to sources of heat;

• all electrical signal and power lines are carried in separate trays;

• all control and solenoid valves fail-safe and, in the case of control valves, their stroke length is correct;

• all ancillary devices are properly installed — for example, alarms, limit switches, air filters and positioners.

It is not necessary to test out every signal and supply line during precommissioning, but in the event of any elements having to be disconnected — maybe because of some other fault that has been found — it is wise to check out the lines when the installation has been restored.

For pneumatic signals and supply lines, instrumentation should be disconnected where necessary to prevent over-ranging, and the lines flow tested for continuity and pressure tested to approximately 2 bar for leakage. All air lines must be clean and dry before being brought into service.

For electrical signal channels and power lines, continuity, earthing and screening should be checked. If necessary, compliance with the requirements for maximum loop resistance should be confirmed; likewise, the minimum requirements for insulation.

6.5 Control system commissioning

The principal objective of the commissioning process is to identify faulty installation and/or operation of the hardware and mistakes in the software. Its scope therefore embraces inaccurate calibrations, incorrectly wired i/o channels, faulty configuration, incomplete operator displays and illogical sequence flow.

The basic strategy for commissioning instrumentation and control systems is the systematic functional testing of all the elements and sub-systems against the functional specification.

The elements referred to may be either hardware, such as transducers and valves, or configurable software functions such as alarm lists, displays face-plates and control blocks.

The sub-systems will be a mixture of hardware and software, varying in complexity from single i/o channels, interlocks, closed loops and sequences through to complex control schemes and self-diagnostics.

Functional testing of i/o channels entails applying test signals and visually checking the response. For inputs, the test signals are simulated process signals, either generated during water or air tests, or applied manually. For outputs, control signals generated by the computer are used.

As a guide, Appendix 6.1 (pages 104–105) provides a check-list of some factors to be considered during the functional testing of a control loop. Similarly, a check-list of some factors to be considered during the functional testing of a sequence is given in Appendix 6.2 (pages 106–107).

In addition to the general purpose control systems, their displays and safety related functions, there are often separate systems dedicated to:

- emergency shutdown (plant protection);
- management information;
- 'packages' — items of plant, such as compressor sets;
- specialized instrumentation, such as chromatographs;
- back-up power/air supply.

From a commissioning point of view, they all consist of similar elements and sub-systems, and the strategy for commissioning them is the same; but they often have different functional specifications and particular care needs to be taken at the interfaces between them.

The early installation of the computer control system enables its use as a sophisticated commissioning tool and can potentially have a major impact on the overall commissioning process — but the impact of this on the schedules for the installation and precommissioning of the field instrumentation needs to be carefully planned.

6.6 Management

Each of the commissioning activities must be broken down into a number of manageable tasks. And for each task a schedule needs to be established with benchmarks for monitoring purposes. The rate of commissioning is measurable (for example, the number of loops/displays or sequence steps tested per day), thereby enabling progress to be reviewed regularly.

For each sub-system successfully tested, the engineer responsible should sign an appropriate test form, preferably accompanied by proof of testing, typically in the form of printout. Examples of various test forms for instrument calibration, alarm system checks, loop test and so on, are given in BS 6739, Reference 6, from which Appendix 6.3.1 (pages 108–109) represents the form for loop testing.

Inevitably, there will have been changes to the functional specification necessitating modification of both configurable and procedural software. Mistakes will also have been revealed during commissioning, so a procedure should be in place to enable software change. Noting the implications on safety, this procedure makes authorization of changes difficult, but their subsequent implementation easy. This strategy ensures that software changes are treated just as seriously as modifications to the plant — for example, for Hazop purposes — but takes advantage of the user friendliness of modern systems.

The procedure must involve the completion of an appropriate modification control form, an example of which is given in Appendix 6.3.2 (page 110) which outlines not only what change is required, but also why it is necessary. The details subsequently become embodied within the documentation.

The position of Hazop studies in the software cycle is indicated in Figure 6.2. An important point to appreciate is that it is the documentation — that is, the software-related parts of the functional specification — which is subject to the Hazop studies and not the software itself! It is against the functional specification that the software is tested and commissioned.

For computer-controlled plant, it is also recommended that at least one 'Chazop' or computer Hazop is carried out. This addresses the hazard potential of the software (and computer hardware) itself. As with the Hazop, the Chazop may also need to be revisited as a result of changes required during commissioning.

Generally, if there is a change in the functional specification, depending on the nature and scope of the change, it is necessary to carry out a further Hazop study before implementing the appropriate software changes. Mistakes in the software revealed during commissioning can, however, be corrected without further Hazop considerations, provided the functional specification has not been changed. On most major plants, there is an independent protection system and changes to the control system software should have limited or minimal impact on the ability to shut the plant down safely.

Some important considerations regarding the modification of safety-related software are given in Check-list no. 16A of the HSE Guidelines, Reference 7.

Functional specification

Hazop

Develop software

Test software

Installation

Commission software

Acceptance

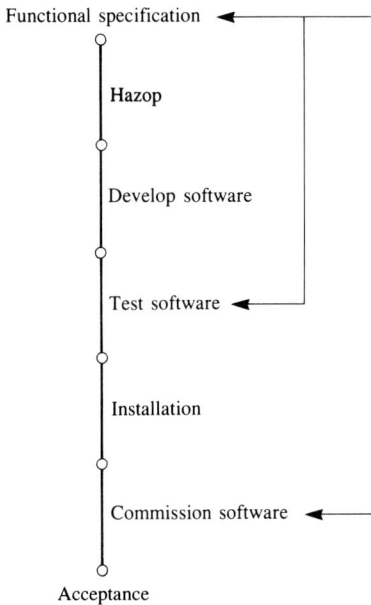

Figure 6.2 The role of Hazop in software development and commissioning

Although implicit in the above, it is nevertheless worth emphasizing that the disks, hard or floppy, and tapes containing the systems software, database and the applications software are themselves covered by the procedure for making software changes. The disks, including back-up disks, should be labelled clearly to indicate which version of software is stored on them. The master is kept separately and not updated until the software test have been fully completed. Software version control is made more secure if issue is made using the medium of CD-Rom.

The segregation policy will have been developed at the design stage. This ensures that the various i/o channels are grouped on the basis of:

- signal type;
- plant item/area;
- process function and so on;

as determined by the requirements for:

- card/rack organization;

- intrinsic safety;
- electrical isolation;
- emergency shutdown;
- manual operation;
- power amplification.

It is essential that the segregation policy is not compromised in the course of making modifications. It is all too easy, for example, to connect up and configure an additional signal using an apparently free channel but overlook the segregation implications.

Access to the system is an important consideration. There are two types:
- access to the software by control engineers for commissioning the system;
- access through the system by process engineers for commissioning the plant.

To accommodate the needs for access, which peak during commissioning, temporary extra VDU/keyboards may be necessary.

Access needs to be strictly controlled. This is normally accomplished by means of key switches and/or passwords, with the levels of access being determined by the nature of function/responsibility/expertise. Passwords and levels need to be recorded and reviewed regularly, with written consent for changes if necessary.

Because of the distance between the various elements of the loops, interface racks and control room, the use of portable phones for two-way communication during testing is essential. This, of course, is subject to intrinsic safety and interference requirements, especially with regard to corruption of computer memory.

6.7 Personnel

Successful commissioning of instrumentation and control systems needs to be considered within the context of the overall commissioning programme. Good planning, coordination, communications, leadership, teamwork and training are all essential.

The commissioning team consists of a mixture of computer and control specialists, instrument and process engineers, the size of the team and its mix obviously depending on the nature and scope of the system. Typical responsibilities are illustrated in Table 6.2.

The applications software is best commissioned by the process engineers who will have responsibility for operating the plant. It should not be commissioned solely by the engineers who developed and tested it. Control system software specialists must be available throughout all phases of commissioning (24-hours a day if the plant is being commissioned around the clock).

Table 6.2 Responsibilities of specialist disciplines in control system commissioning

	Instrument	Control	Process	Computer
Instrumentation	✓			
Field wiring	✓			
i/o system	✓	✓		
Computer h/w	✓	✓		
Systems s/w		✓		✓
Database		✓	✓	
Configurable s/w		✓	✓	
Procedural s/w		✓	✓	

The automation function very often becomes the lead discipline upon which the progress of other disciplines depends. It is important therefore to have adequate resourcing at the correct level to handle the inevitable peak demands.

It is normal practice for the system supplier to provide specialist support *pro rata* during site commissioning; it would be contractually difficult to arrange otherwise, there being so many factors beyond the supplier's control that influence the time and cost involved.

There is much to be gained from involving the operators in the commissioning process:

- it familiarizes them with the system and, as such, constitutes training;
- they can contribute to the modifications — for example, mimic diagram layout and warning/alarm settings.

6.8 Documentation

For turnkey systems, once the supplier has been selected, the scope of supply defined and a detailed applications study completed, all aspects of software and hardware functionality are pulled together into a single document: the functional specification. This becomes the reference source for software development and testing. In particular, it should contain agreed and comprehensive test criteria for all applications software, against which works testing and site commissioning can be carried out (see Figure 6.2).

This process of documentation for software is discussed in detail in the Institution of Electrical Engineer's (IEE) Guidelines, Reference 8. By virtue of this process, applications software may be produced and commissioned in accordance with Quality Assurance criteria (Reference 4).

The documentation associated with instrumentation and control systems can be extensive and includes:

- process and instrumentation diagrams;
- loop diagrams;
- database tables;
- procedural coding;
- wiring/circuit diagrams;
- termination rack layouts;
- tag number reference lists;
- systems manuals;
- configuration charts;
- sequence flow diagrams and so on.

All involved in commissioning have a responsibility to identify any inconsistencies in the documentation and to ensure that they are rectified. It is particularly important that modifications made during commissioning should be incorporated, and it is clearly necessary to have a procedure to ensure that this takes place (see Appendix 6.3.2, page 110).

Process commissioning

7

Process commissioning begins when the precommissioning tests have been completed, and faults corrected to the satisfaction of the commissioning manager. It is when plant, equipment, systems or sub-systems are put into operation for their normal duty. This is the time when the thoroughness of the planning and preparations described in the previous chapters is put to the final test. It cannot be too strongly emphasized that there are no shortcuts to following those procedures, no matter how large or small, complex or simple the project.

As discrete systems or sub-systems become available after completion of the precommissioning tests, and faults have been corrected, it is often possible to commission them on process fluid without interfering with outstanding construction or testing work, using temporary systems or equipment if necessary. Because of the potential dangers associated with having some parts of the plant running while the overall plant is unfinished, the decision to do this should be taken at a senior level. Obviously clear instructions must be given to all construction and commissioning staff to ensure they are aware of the up-to-date situation. It is good practice for the commissioning manager to receive a formal handover certificate from the manager responsible for mechanical completion, verifying completion and listing minor deficiencies.

Where the site or plant is registered to the Quality System BS EN ISO 9000, it is important to ensure that the requirements of the quality manual are properly carried out as commissioning proceeds — see Chapter 2 (Reference 4).

7.1 Final checks

Particular emphasis must be placed on safety awareness throughout the commissioning team to run through a final check-list of items which may cause difficulties if not available when required. Such a check-list would include:

• all safety equipment such as safety showers, eyewash bottles, breathing sets, and fire extinguishers in place and functional, emergency procedures known (see Chapter 4) and medical/first aid arrangements complete;

- all inspection equipment removed (such as ladders from vessels) or disconnected;
- test orifices and blanks removed (especially from vents and overflows);
- strainers cleaned and replaced;
- temporary screens to protect pumps and valves in place where necessary;
- liquid seals in vents and manometers at correct levels;
- emergency and normal lighting systems functional;
- rotating equipment guards in place;
- all sources of ignition removed (for plants handling flammable chemicals);
- relief systems fitted;
- access for emergency vehicles not obstructed;
- tank/vessel bunds clean and empty;
- laboratory prepared to receive samples;
- all temporary equipment, consumables and discharge routes are available;
- all documentation available in the plant such as calibration charts, log sheets, process manuals and data sheets, handover logs, operator training log and quality manual.

When these checks have been done, the commissioning manager usually calls the team of operating staff together to explain the sequence to be followed and to allocate specific duties. Communication between individuals and the manager is vitally important during this phase of the project.

7.2 Typical sequence of commissioning

Depending on whether the plant is on a greenfield site, an extension to an existing complex, or a revamp of an existing plant, a typical sequence of commissioning may be:

Utilities
- electricity, including emergency generators;
- water and its treatment systems;
- compressed air for instruments, process and breathing;
- demineralized water;
- natural gas receiving station;
- fire-fighting water;
- steam boilers and economizers;
- condensate;
- nitrogen or other inert gas from cylinders or a production unit. These must be regarded as process chemicals and blanked off from process equipment until otherwise authorized.

- refrigeration;
- warm or pressurized water systems;
- drains and effluent treatment systems.

Laboratory

Setting up a laboratory is outside the scope of this guide. Suffice it to say that a laboratory is a key element which must be thoroughly prepared and ready to deal competently, reliably and accurately with all anticipated samples before attempting to start the plant on process fluid.

Adequate sampling points are provided on the plant and sampling procedures reviewed for safety and operability before the start of commissioning.

Storage tanks/silos for raw material

This also includes tanks and silos required to service utility plants — for example, fuel and diesel oil, liquid nitrogen or other inert gas and LPG.

Ancillary equipment

Many plants have substantial systems which are ancillary to the primary function of the plant. Examples are:

- drum melters;
- ventilation and fume extraction systems;
- scrubbers and absorbers;
- pressurized/warm water systems;
- vaporizers;
- vacuum systems including phase separators, distillate receivers and solvent recovery;
- intermediate storage tanks/vessels.

It is often possible to commission these systems progressively as they are completed and precommissioned.

Reaction

The manner in which the plant is commissioned will depend largely on whether it is batch or continuous. Sometimes plants can be divided into several discrete stages where intermediates are made continuously then tested and stored for short periods before proceeding to the next stage. Commissioning of a reaction system is dealt with in more detail in Section 7.7 of this Chapter (page 50).

Reaction mixture work-up

The majority of plants require further process stages subsequent to reaction. This equipment must be tested and ready to receive reaction mixture as and when it is available from the reactor. Such equipment may include fractionators, evaporators, solvent extractors, crystallizers, neutralizers, filters and blenders. It is sometimes advantageous to commission downstream units by the preparation of a synthetic mix or by importing intermediates from another plant.

Plants without reactors

Some process plants do not involve reactions, so items in the preceding two paragraphs may not apply. In these cases, raw materials may be processed in equipment such as blenders, driers or kilns, pastillators, grinders and granulators.

Product storage

Product may be stored in various forms, such as gas, liquefied gas, paste, slurry, powder and pellets. Storage may be for intermediate products, recycle streams, by-products or finished product. The time at which they are commissioned and need to be available is determined by the stage of the process they fit into. It is essential, therefore, that the precommissioning tests are done to a carefully planned schedule.

7.3 Quality control laboratory

Before attempting to start the plant, all the in-process quality checks must have been defined, methods agreed for various analyses and what corrective action is to be taken if a result is outside the desired range. Equally important is to check that laboratory staff have been trained and are kept advised about when to expect the first sample from the plant. If there is serious doubt about any aspect of quality control, it should be cleared up before the plant is started up, in just the same way as if it were a problem with a piece of process equipment.

Key aspects of quality control to be checked include:
- hazard potential and treatment data for all material handled;
- specification, methods of sampling, equipment for sampling, methods of analysis and acceptance tests for raw materials, process intermediates and finished products;
- equipment calibrations;
- all methods of analysis which should have been agreed with supplier, client or customer as appropriate.

A systematic method of recording times and stages of the process from where samples were taken, along with their results, is an essential requirement of the start-up paperwork. Often these results can provide useful clues to solving unexpected problems which might arise, perhaps from inaccurate instruments. During commissioning it is inevitable that a larger number of samples will be analysed to check the plant is performing correctly than is the case for routine operation. The laboratory management needs to be aware of this to arrange staffing to suit.

7.4 Ancillary equipment

The process flow-sheet is checked to identify specific pieces of plant equipment which either:

- must be commissioned and operating satisfactorily before feedstocks are introduced to the reactor — for example, a scrubber may be essential for cleaning a vent gas before discharge to atmosphere; or
- can be commissioned on chemicals before starting a reaction proper.

An example would be the wash, separation and filtration stages of a plant which is an expansion of an existing process. In this case it may be possible to take from the existing reactor product, which is known to be of good quality, and work it up in the new plant.

It is wise to commission and gain confidence in the operation of ancillary equipment before starting up a critical stage of the process. This allows effort to be concentrated on the most difficult stage of the process without diversions on trivial faults with ancillary equipment. In the example just described, the scrubber could be filled with the specified liquid and the pump run for several days before required by the process. Any leaks or pump seal problems can then be remedied in good time.

Once it is decided which items fall into these two categories, a list of priorities is drawn up. Where the process involves expensive or hazardous chemicals, consideration should be given to initiating commissioning with relatively cheap chemicals possessing similar properties.

7.5 Storage – raw materials

Consideration must be given to the availability of raw material both in drums/packages and bulk. Close liaison with the purchasing department needs to be maintained to ensure that delivery schedules correspond to the commissioning programme.

- drums — where high melting or high viscosity raw material must be heated to charge to the process, drum heating appliances need to be tested and commissioned well in advance of start-up. Temperature controllers and alarms must function correctly. Obvious items such as checking that the correct weight and grade of raw material has been delivered and sampled to ensure that it meets the purchasing specification, can easily be overlooked;
- packaged dry goods — store under cover and acceptance test as required;
- bulk containers — as above with additional checks to ensure that the containers are as specified and will connect properly to the plant. For greenfield sites, also ensure that fork lift trucks are available to receive and locate the size of containers specified and that they have adequate reach and weight capacity;
- bulk — raw material storage plants are usually commissioned in good time though the cost of storing expensive materials for unnecessarily long periods must be borne in mind.

Potentially hazardous raw material tanks should receive special attention from an experienced commissioning engineer before and during the off-loading of the first tanker. A typical list of items to be checked before discharging into a new tank is given in Appendix 7 (page 111).

If all the careful testing and preparation has been carried out thoroughly as described so far in this guide then unloading the first consignment should go smoothly and soon become routine. If something does go wrong, the previously thought-out emergency procedures will bring the situation quickly under control.

7.6 Storage – intermediates and finished product

Before commissioning the reaction system, any intermediate and finished product tanks must have been checked in a similar way as for raw materials — see Appendix 7. Additional items to check are the transfer systems between tanks and road/rail tankers.

Sometimes during start-up, material from the process or from a part of the process may be of such poor quality that it cannot be further processed or recycled. Careful thought should be given before the start of commissioning to disposal routes and arrangements for such out-of-specification material and discussed with the Environment Agency.

7.7 Reaction system

Only when the commissioning manager is fully satisfied that the steps described so far have been completed and all equipment commissioned to an acceptable level of reliability, should any attempt be made to commission a reaction system. Because of the wide range of degrees of complexity of reaction systems, they are sub-divided into continuous and batch and general guidance is given on typical commissioning procedures for both.

Continuous plants

Continuous plants are usually characterized by high throughputs, and fast reactions with relatively low inventories of reactants under closely-controlled parameters of flow rates, temperature and pressure. Modern plants generally have computer or PLC process control. The commissioning manager's main concern at start-up is to achieve stable process conditions as quickly as possible and then to be confident of achieving a reasonable running time without breakdown. The commissioning manager will be particularly keen to have confidence in instrument calibrations, software programmes and reliability of prime movers. A typical last minute check-list, additional to that given in Section 7.1, includes:

- raw material feed tanks at correct temperatures;
- warm-up routines complete where necessary — for example, steam or electric tracing on feed and product lines and process equipment;
- establishing inert atmospheres;
- ancillary systems operating normally;
- cooling water and emergency systems for bringing potentially hazardous reactions under control are immediately operational if required;
- emergency shutdown procedures are clear and understood by operational staff;
- additional support staff are available as required including process chemists, shift technicians, electricians and specialist fitters.

Once the commissioning manager is satisfied that these checks are acceptable, the precise time to start the reaction can then be chosen. Because the hours, days and maybe weeks following start-up are likely to involve intense mental, physical and intellectual strain on the commissioning team, the commissioning manager will want to ensure that its members are fresh and alert. There are no hard and fast rules about the best time to start-up as it depends on local circumstances, organization and complexity of plant, but some points to consider are:

- if starting up on nights, weekends or close to public holidays, are critical equipment service engineers available at short notice?
- shift change and break times are not appropriate;
- availability of key personnel for consultation if needed.

Start-ups naturally attract much interest and concern of senior management. It is important to curb their desire to be present to 'see how things are going' during this period of intense activity. One way of doing this is for the commissioning manager to give a progress report at agreed times to one nominated senior manager.

The commissioning team often has to decide whether to continue running a plant in what are normally unacceptable conditions, or whether to shut down to correct a fault. Clearly, this depends on the circumstances but it is only possible to learn the characteristics of a continuous plant whilst it is running. It is often better to keep running to establish that the process is producing the right quality product before shutting down. This reduces repeated start-ups, generally producing off-specification products, which may be difficult or expensive to recover, blend away or dispose of.

It is important throughout this commissioning stage to record as much relevant data as possible to check design assumptions and to keep a log of all faults. When the plant is shut down for operational reasons an attempt should be made to correct all known faults.

Batch plants

Batch plants are usually characterized by lower throughputs, slower reactions with relatively large inventories of reactants and multistage reactions or process steps. They may involve multiphase stages, significant changes in physical properties, reaction with or without distillation and so on. Batch plants are often designed to make a wide variety of products. Modern batch plant generally has computer control of process parameters and computer or PLC sequence control of individual process steps and alarm handling. Unlike continuous processes batch ones are always transient and rarely have long periods at stable conditions. They therefore require constant attention, especially at start-up.

Commissioning a batch plant is generally less demanding in respect of engineering and time pressure because the processes can usually be sub-divided into discrete stages, with quality checks being made between each of them. There is greater emphasis on charging accuracy, however, where concentration effects and contamination can cause undesirable by-products, resulting in loss of yield and increased operating costs. Hence the commissioning manager concentrates on ensuring that correct charges, exact process sequences and critical process parameter controls are achieved on the first batch of each product to be made.

For exothermic batch reactions, the heat release is often controlled within the reactor cooling capacity by limiting the addition rate, or weight added, of one of the reactants. A prime objective of commissioning is to confirm the process design and to establish a reliable, safe operating procedure for routine production.

Batch plants are more flexible than continuous plants and so process conditions and sequences can more easily be altered both by design and error! A systematic record of process conditions for each batch is essential so that they can be related to quality analysis results. Such changes in process conditions are formally agreed with the process chemist, and authorized only after a thorough safety/hazard check by the commissioning manager.

Batch plant efficiency, or plant utilization, depends on achieving design batch times and batch outputs. This is equivalent to the throughput rate of a continuous plant. It is usual to concentrate initially on establishing that the process and plant are capable of meeting the product specification. Effort then moves towards achieving design yields and batch times consistently.

7.8 Completion of commissioning

In Quality Assurance terminology, the commissioning stage is not complete until it has been demonstrated to the customer that the plant is 'fit for purpose' for which it was designed. For in-house projects, the customer may be the plant, production or works manager; for large turnkey plants, it is the client company's representative. The 'purpose' may be defined in many clauses in the contract, but is generally stipulated in terms of a product range of specified quality, yield and production capacity. Hence, the customer will require objective evidence that these have been achieved.

For projects where separate commissioning and operating teams are involved, a formal handover from the commissioning manager to the operation manager is essential. The handover must list all items which need continuing technical input to deal with any unresolved problems or difficulties and define who is responsible for completing them to the satisfaction of the customer.

For large turnkey projects this is usually provided by guarantee or acceptance runs which are described in Chapter 8.

Performance and acceptance tests

8

Performance tests are carried out with the plant under design conditions — that is, tested at the design rate and yield using raw materials of specified quality to produce on-specification product.

If the plant or process has been licensed, the licence often guarantees the process. Guarantee tests are supervised by the licensor (or the contractor that engineered the plant) with the plant operated under specified conditions. These conditions are generally outlined in the contract, but often the actual method of operation and specification of measurement techniques are arrived at by joint agreement between the licensor and the operating personnel. If the plant is 'own design and engineering', then an accurate assessment of plant performance at design conditions is required before start-up can be said to be complete.

Ideally, the schedule of performance tests is drawn up well in advance of the actual tests. The duration of tests, performance criteria, conditions governing interruptions, methods of measurement and analysis and allowable tolerances all need to be specified. Methods of determining stock levels at start and finish, recording of instrument readings and sample collection are all included. Some of these requirements affect the detailed design of the plant and hence must be considered during the plant design stage, such as the provision of adequate flow monitoring devices of the required accuracy.

Performance tests usually include plant capacity, consumption of raw materials, and quality of finished product. In some instances the tests may also cover reliability, consumption of utilities, and quantity/composition of effluent discharged.

Conditions of Contract for Process Plant (better known as the IChemE Red, Green and Yellow Books) (References 1, 2 and 3) states that if the contractor guarantees the performance of the plant will there be a schedule of performance tests and that failure of the plant to pass these tests shall render the contractor liable to pay liquidated damages. Further details of the Conditions of Contract for performance tests are given in those references.

If these performance conditions are not laid down in the contract then they should be set by the plant management.

A performance test may be carried out when the conditions presented in the contract have been attained. These may include:

- the plant operates under steady controlled conditions — that is, all temperatures, pressures, flows, levels and analyses have been fairly constant (or repetitive in the case of a batch process);
- product specifications are being consistently achieved;
- mass and energy balances have been established and show agreement both weekly and daily.

When all these conditions have been met, the operating conditions for the performance run must be established and maintained for a time period before starting the performance test, (this operating period may include the performance test). During the performance test, data are recorded accurately and all pertinent readings verified jointly by representatives of the guarantor and the plant management.

When the data have been collected and processed, corrections are sometimes necessary to allow for such things as actual cooling water temperature versus design temperature and heat exchanger fouling.

There are three possible outcomes from the performance test for the plant:

- it achieves the design output and efficiency;
- it does not achieve either output or efficiency but the reason can be identified and the problem overcome. Here, the contractor may either put the problem right at the contractor's expense, or alternatively pay the purchaser an agreed sum so that the problem can be rectified at the next convenient plant shutdown;
- it does not achieve either output or efficiency and, due to design errors, it is unlikely that it ever will. In this case, the contractor will have some financial liability and the purchaser will suffer consequential losses.

Postcommissioning documentation

9

The changes and modifications to process plant that occur as a result of commissioning experience are the major source of operating hazard.

Pipework may be modified, valve operation totally changed, reactors operated in a manner significantly different from the design concept, and many small changes made to the design as flow-sheeted. Sometimes these changes are not properly incorporated in the documentation, so that the risk of misunderstandings over the life of the plant is increased. Systems must be in place, and followed, to ensure that all changes are incorporated into 'as-built' documentation. Temporary modifications are also common during the early operational phases; these also bring risks. No modification should ever be made without a careful appraisal of the implications and reference to the design intent. These aspects are covered in Chapter 3.

9.1 Modifications

Any modifications that affect standard operating procedures, drawings, plant troubleshooting, plant numbering, equipment schedules and maintenance need to be considered by the commissioning team. If they have not been covered in the contract, they are the responsibility of the owner. If serious changes have occurred a Hazop or design review study is essential and, if necessary, expert advice should be taken.

9.2 Records

All relevant records from performance tests, insurance reports and first production runs should be incorporated in the documentation system to act as datum points for long-term operation.

9.3 Computer records

Any computer records — for example, control setting changes and other changes to software — must be recorded and if necessary audited (see Chapter 6).

9.4 Audits/reviews

A postcommissioning review should be held and information fed back to commercial departments, designers and contractors. If necessary, some aspects should be subjected to an independent audit.

During the first few months of operation of a new plant, the reliability and operability usually improve rapidly. It is important that this experience is fed back to the design staff and where appropriate the equipment suppliers advised of shortcomings in their designs.

9.5 Reservation lists

The reservation list concerning defects noted during commissioning should be processed and passed on to operating management. This is essential if rectification must be deferred to a shutdown or otherwise fitted into operating practice.

9.6 Maintenance

Commissioning spares should be replaced with the appropriate spares holding and any commissioning oilfills should be flushed and replaced. Full spares inventory procedures may be needed if the project is large enough. A list of service department addresses and telephone numbers should be made available to operators.

9.7 Logsheets

A developed logsheet and records procedure is an essential component of safe and reliable operation.

Special cases

10

10.1 Small plant commissioning

Small plants are affected by all the normal problems associated with major projects such as late drawing office issues, late manufacture, design errors, late deliveries, drawing errors, installation problems and so on. These problems can affect the commissioning activity. The only thing which rarely changes is the completion date. Consequently, the time allocated for commissioning often ends up being half as long as it should be with twice the work content.

On major projects, spread over a comparatively long time-scale, there are options to absorb much, if not all, of this slippage. But many contracts do not fall within this category; a commissioning period of one to six man weeks is often the norm and carried out by one multiskilled engineer.

Because most of these types of plant are to be installed in an existing complex, many of the plant services already exist and have only to be extended to termination points at the new equipment, able to provide the correct rate, pressure and temperature.

The plant is installed by skilled craftsmen working to plant layout and isometric drawings. The only involvement by a commissioning engineer at this stage may well be a short visit to inspect the installation for familiarization and checking that the drawings are being correctly interpreted. If the control system is micro-processor controlled, the commissioning engineer may well have already been involved in a demonstration at the supplier's premises. The engineer will also have spent time collating the necessary drawings, line diagrams, data relating to bought-out equipment, process description, contractual documentation and preparing the necessary check-sheets which will form the historical data.

The preparation work for contracts of this type is at least as important as the implementation. The job will only be as good as the forethought which has gone into it.

Creating and updating 'as-built and commissioned' documentation is one of the most important aspects of commissioning. A daily log of events should always be kept. Such details as proportional integral derivative terms, timer settings, purge times and sterilizing temperatures, should be recorded on the appropriate forms, together with a marked-up copy of software listing changes and wiring modifications.

The client's personnel are encouraged to become involved with the project at the earliest opportunity, both from an operational and a fault-finding point of view. That way, when the 'crutch' — in the form of the commissioning engineer — is removed they will not all fall flat on their faces.

10.2 Decommissioning

In projects where an existing plant is being modified, extended or replaced, part of commissioning the new plant may involve closing down and removing existing plant, or modifying existing processes. This is fully covered in the IChemE sister text entitled *Decommissioning, Mothballing and Revamping* (Reference 10).

Features that are particularly associated with this phase of commissioning are:

Records

Records of electrical supplies, drainage routes, vessel contents and structural design may be inaccurate or even non-existent. There is a considerable risk to decommissioning staff and the security of adjacent operating plant during this phase due to error in records.

Area classification

The area classification may change while decommissioning is occurring. If plant is to be scrapped, the contractor's personnel must be carefully supervised for their own safety as well as that of the adjacent plant.

Safety

The presence of asbestos, flammable materials, toxic and other hazardous substances may require special consideration in order that the new plant can be removed safely and suitable procedures must be devised to cover all aspects.

Venting and purging

Cleaning out vessels and lines that have been in use usually involves venting, purging and pacifying. These activities are carefully planned with attention given to pollution and firerisk. For example, the Environment Agency monitors and agrees procedures where river and waterway pollution is possible.

10.3 Projects carried out abroad

Much of the content of this guide is as applicable to projects carried out abroad as to those in the UK. The former frequently present particular difficulties, and failure to recognize them at the outset can adversely and sometimes seriously affect the success of a project. Some aspects which come within this category are referred to under specific subject headings in the main text. This section seeks to provide a brief summary of some of the more important features affecting commissioning outside the UK.

One potential difficulty in some locations is that of protracted timescales. This possibility needs to be considered at the bid preparation and contract negotiation stage, having regard to commissioning as the final phase of the project and thus likely to be most affected.

Aspects likely to have particular significance in contract conditions and associated terms and conditions relating to commissioning include:

- definition of erection completion and protocol;
- client operator training 'abroad' and at site;
- numbers of commissioning personnel including specialists and method of payment;
- duties of commissioning team, specialists and other experts. Responsibility for instructions; routing of directions, usually via the client's technical personnel; recognition that physical operations are performed by the client's personnel under supervision of the client's technical staff;
- log book/log sheet and operating advice sheet completion; dual/multi-lingual if appropriate;
- provision of interpreters adequate in number (to include shift duties);
- continuity of services, particularly power supplies and adequacy of emergency supply for essential services;
- provision of laboratory services;
- interruptions to commissioning period;
- effect of local religious observances and similar;
- definition of performance demonstration and guarantee test; interruptions to test runs;
- definition of plant acceptance and protocol;
- provision of office and communication facilities; the latter should include availability at times when home offices/project personnel can be contacted.

A most important corollary relates to the welfare and morale of the commissioning personnel to be deployed and the selection of the team itself. Commissioning in a foreign location can put enormous strain on individuals and particularly on the resident team leader who requires all-round capability in addition to technical competence. Compatibility of the team, particularly under stressful conditions, is also an essential requirement. Terms and conditions for assigned personnel need to take into account such matters as:

- inducements such as monetary incentive and sometimes assignment completion bonus;
- assignment status, (single/married and accompanying children); schooling;
- provision of air fares and excess baggage;
- accommodation and furnishings;
- local currency provision and banking arrangements (if any);
- availability of food supplies;
- local transport including the use of vehicles for recreational purposes;
- medical facilities;
- leave frequencies; local and home;
- behavioural importance, off-duty as well as on-duty;
- compliance with local laws and regulations;
- repatriation arrangements.

It is vital that the commissioning team members are adequately briefed on the project before taking up a foreign assignment; similarly that they can be assured of prompt response to communications from site since failure to do so can have a very damaging effect upon morale.

Staff assigned should also be familiar with procedures to be followed in the event of national emergencies or other circumstances which might require evacuation of ex-patriot personnel.

10.4 Bioprocess plants

10.4.1 Introduction

Processes which utilize micro-organisms to effect conversions and those which further transform materials arising from such biological activity into commercial products have characteristics which set them apart from conventional chemical process operations. Generally, metabolic processes occur in dilute aqueous media under mild conditions of temperature, pressure and chemical agressivity and require small energy levels to promote the catalytic action of enzymes. These process conditions must be carefully controlled in order to maintain cost-effective levels of process intensity and conversion efficiency. It is the imposition of such controls (as well as the need to ensure safety, integrity of operation and regulatory approvals) that has most impact on bioprocess plant design and commissioning.

Before defining the necessary differences from commissioning conventional plant (as set out in the preceding sections of this guide), however, it is useful to appreciate the range of scale and technology of commercial bioprocess plants. Some biologically-based industries such as bread, wine and sewage treatment have their origins in prehistory and these remain the largest (volumetrically) process operations with bioreactors of up to 1000 m^3 processing 10,000 m^3/day of broth. More recently, processes have been developed in which genetically manipulated organisms are grown and purified, generally on a smaller scale. For some products, 'laboratory-scale' equipment is sufficiently large to satisfy market demands. The approach to engineering design varies depending on the product. For example, there is little need for sterility, monosepsis and containment at the large end of the scale but almost obsessional care is taken when dealing with active viruses and hormones.

Biological processes tend to be batch operated with various control conditions that are changed over the course of one to ten days. They utilize natural feedstocks which are also subject to considerable variation. Fermentation broths are complex three-phase mixtures which although they may have non-Newtonian rheology, must be well agitated for mixing, heat transfer and mass transfer. Often the concentration of biomass and product cannot be directly measured but must be inferred.

The product may be biomass itself, an organic molecule which is an excreted secondary metabolite or a complex protein which is accumulated within parts of the microbial cells. Consequently there are many ways by which products are extracted and purified. However, water removal, biomass removal and further separations based on chemical and ionic properties are usually employed to produce a crude product which may be further refined and subjected to chemical conditioning before the final product is realized.

Biological processes are very different from conventional chemical reactions which obey well-defined laws of chemistry, reaction kinetics, thermodynamics and catalysis. In biological processes, living cells must be supplied with nutrients and oxygen and as they grow they can be conditioned to effect the required molecular transformations. The approach to bioprocess plant commissioning is in many ways different from the guidelines provided in the preceding sections.

10.4.2 Safety

High temperatures and pressures, aggressive reagents and hazards associated with runaway reactions are not generally present on bioprocess plants. The most prevalent physical hazard arises from the inadvertent escape of sterilizing steam and simple precautions can effectively reduce such risks.

The real potential hazards relate to process operations involving biologically active materials which may be invasively hostile, infective and cause physiological damage. These hazards are not usually present during the commissioning phase of a project but must be anticipated when, after handover, the operating company introduce, live micro-organisms. It is the commissioning and operating engineers' task to ensure that the plant is rendered completely safe for bioprocess operation.

In order to do this, the commissioning and operating engineers are involved in projects at an early stage. They need to become conversant with the process technology when the process design is being executed and take part in safety reviews.

Process operator safety, environmental safety and product safety are interrelated and the preparation of documentation which provides the basis for establishing formally that safe conditions prevail also includes the preparation of very detailed precommissioning, commissioning and operating manuals. The handling of biologically hazardous materials is subject to nationally recognized guidelines and approvals legislation. It is necessary to prove that the design and operation of such plant ensures total containment of viable organisms and potent materials with in-built safeguards to deactivate them before release from the contained process system. It is also necessary to provide documentation concerning the testing of the plant to demonstrate that under all possible normal and fault conditions, the safety integrity of the plant is not breached.

10.4.3 Quality Assurance

In order to comply with legislative requirements, a new pharmaceutical facility has to be granted a manufacturing licence before it is permitted to produce a recognized product for which a product licence has been or may be granted. The submission of documentation for approval and the demonstration of satisfactory performance is called validation and this can only be effectively achieved if this is fully recognized at the initiation of the design, engineering and construction project.

Validation entails the systematic proof, by means of witnessed stage-by-stage checkouts, that the process plant, infrastructure and building system meet the approval of the Medicine Inspectorate or other legislative authorities such as the American Food and Drugs Administration (FDA). It is necessary to demonstrate that Systems of Work and Standard Operating Procedures (SOPs) comply with the approved Good Manufacturing Practice (GMP) for the product and that product quality is consistently in compliance with specification for quality and therapeutic effectiveness. This can only be achieved once a comprehensive Quality Assurance regime (including engineering records and testing) has been instituted.

The operating engineer is a key member of the project team and is instrumental in the interpretation of process technology into detailed instructions for plant operations. Because of the special requirements for pharmaceutical plant, it is recommended that at least one engineer is appointed to this task at the start of engineering projects for technical liaison with plant operating company personnel who are normally responsible for seeking statutory approvals. This engineer should also have a special responsibility to provide the preparation of functional test procedures which establish validatory compliance for each equipment item or operating system by bringing together complete design and manufacturing documentation with precommissioning test procedures.

Whenever possible, a systems approach is adopted which provides consistent documentation standards and procedures for every system and includes checklists, summary sheets and referencing guides that interrelate equipment manufacturer's data with the relevant stages of validation.

10.4.4 Commissioning activities

Many of the unit operations employed in bioprocess plants have been recently developed from laboratory-scale research work and often their pedigree is evident. The commissioning engineer can provide valuable advice concerning design for efficient commercial-scale operation if involved at an early stage. Unfortunately, there is seldom sufficient time during the course of a project to develop more effective scaled-up process systems and so the commissioning engineer must impose adaptations to operating procedures to account for design imperfections. None of the unit operations used in bioprocessing, however, are unique to this industry and whatever scale of operations is required, other non-biological applications of, for instance, chromatographic, ion exchange and ultrafiltration separation systems can be found.

Precommissioning bioprocess systems is similar to the normal routine practised in the chemical industry. There then follows a series of special procedures which ensure that effective hygienic and monoseptic operation can be achieved. Complex sequences of washing with surfactants and caustic reagents (or occasionally dilute acid) are followed by rinsing, displacement of air by stream, stream sterilization and then sterile cooling.

Validated plant is subject to a further series of tests which include challenging the entire process envelope with a simulated external infection and internal process culture in order to prove the integrity and containment.

Utility systems such as water for injection (WFI), clean steam, clean-in-place (CIP) solutions and sterile process air must be similarly proven. Also the building system itself has to be validated. Many bioprocess operations which contain potentially hazardous materials are operated in closely-controlled negative pressure enclosures with filtration of exhaust ventilating air. Sterile and particularly parenteral products are processed in clean rooms which are maintained at positive pressure with filtered incoming air. Validation of building control systems and of personnel changing facilities and systems of work are necessary to meet GMP requirements. Manuals for formal test procedures are required to validate these activities.

Very few sensors are specific to the complex compounds present in bioprocesses. Sensors need to be non-intrusive so as not to jeopardize containment or sterility; because of this many process parameters are inferred. Although flow, pressure and temperature can be measured and sterile probes for pH and dissolved oxygen tension (DOT) are used, there is less process information available than is common in chemical plant control systems. This imposes particular problems during commissioning and initial stages of process operation, and much skill is required in the interpretation of operating data. This emphasizes the need for commissioning and operating engineers to have a detailed understanding of process technology and the plant design, and to have been involved in engineering projects from their earliest stages.

Another crucial area in bioprocess plants is the validation of control systems. Instrumentation calibration must be followed by rigorous checking of control system software. Many biotechnology plants are capable of multiproduct operation and it is essential to ensure that there is no possibility for commonality in product-specific software so that the integrity of GMP for one product does not corrupt a system that has already been validated.

10.5 Secondary pharmaceutical plant

10.5.1 Introduction

The main differences between the commissioning of a secondary pharmaceutical plant and the commissioning of a chemical plant is that the pharmaceutical plant has to be very clean, particularly if it is to be used for the manufacture of sterile or aseptic products, and that a validation procedure is added to the standard commissioning phase.

Bulk pharmaceuticals and generally the active ingredient in a pharmaceutical product, are more often than not produced in primary pharmaceutical plants which are very similar to fine chemicals plants. They are capital intensive, are relatively large and work on a three-shift system. There is a tendency for pharmaceutical companies to locate their primary plants near to their home base. On the other hand, secondary pharmaceutical plants process the ingredients of the product, and fill and pack the medicinal item. These plants are not so capital intensive and work mainly on day shift.

Pharmaceutical companies have high research and development costs to bear. In order to provide these funds, they have to sell their products worldwide, and not just in their home country. Countries often demand that if a pharmaceutical is to be sold in their country, then the company concerned must conduct some manufacture in that country. The result of this is that secondary manufacturing facilities are set up in many foreign countries, with the result that the secondary commissioning activities of large pharmaceutical companies are often done abroad. This is an important factor to bear in mind, as up-to-date knowledge of the construction and operation of pharmaceutical plants with 'clean room' facilities is not present in all foreign countries.

10.5.2 Commissioning of 'clean rooms'

'Clean rooms' normally have autoclaves and ovens set into the walls to allow materials to be sterilized on their way into the room, and to provide a terminal sterilization on the way out. There is normally a sequence of two sets of double doors for large equipment to be brought in and out and other doors into staff changing rooms. The floor, walls and ceiling are of a smooth construction with minimum fittings on the surface and are non-porous and easy to clean. The doors are often polished stainless steel; air enters the room from diffusers in the ceiling behind which are terminal filters, and air is withdrawn through floor level grids in the walls. Light fittings and windows are flush fitting on the inside. Services probably enter by the ceiling above the floor-mounted equipment.

Precommissioning and the rough cleaning of the room can be done by the contractor or the construction team. Final cleaning is part of commissioning and is done by the commissioning team and the operating production staff. Before final cleaning, the surfaces are inspected in detail; if part of the floor is found to be in need of regrinding later, a complete clean will be redone, adding significantly to the commissioning time, so a careful inspection pays off.

When commissioning air handling plant, the main filters are checked for an even passage of air which denoted absence of flaws or damage holes. The fault most generally found is where the high efficiency particle arrestor (HEPA) filters do not fit properly into the frames; there is no substitute for strong frames. The shipping and storage of HEPA filters is most critical. Plenty of spare filters should be ordered and the commissioning team would do well to supervise the off-loading of filters on site and to ensure they are kept in a special store, possibly under the control of the commissioning team. Filter testing equipment is expensive and the testing is a specialist job. It often pays to employ a specialist to fit and test the filter, even if the team has to be imported from abroad. During the commissioning of the air handling systems, the prefilters, which take the heavy and coarse particles out to prolong the life of the expensive main filters, may become clogged and have to be changed. The commissioning team must ensure that these spares are also available. Terminal filters are fitted onto the end of each duct entering the clean room and these should also be tested.

A suite of clean rooms and changing rooms, laundry and so on, have planned pressure differentials between them. These differentials have to be set and the whole system finally checked with the automatic stand-by fans brought in on simulated fan failure while the room pressure differentials are carefully watched to ensure they remain within specification. Controls may have to be adjusted to bring them into line. Fan failure can occur when a door is open, thereby increasing the necessary quick stand-by response. This situation must be able to be handled by the system.

The final cleaning of the room is done with disinfectant prior to a general disinfection of the whole ventilation system. Changing rooms and laundry must be working before the final clean which is done with staff fully garmented. Meanwhile, autoclaves and ovens are commissioned and the floor equipment is wiped down, sited connected to services and covered while final cleaning takes place.

Regular monitoring of the particle count in the room takes place while the air handling system is continuously running.

For a greenfield site the quality control, analytical laboratory and microbiological laboratory have to be fully proven before main plant commissioning takes place. The precommissioning should be done by the future laboratory staff, assisted where necessary by the commissioning team. In a foreign location this calls for early recruitment of staff and their training.

The standard of cleanliness of clean rooms is set down in US and British Standards. Most of the problems in clean rooms come from people and their clothing and so close attention has to be paid to these factors.

After the room and its services are commissioned, the product can be introduced. Thorough checking of the product is essential, particularly where sterilization would harm the product and that product has to be produced and packaged aseptically. Naturally, the points where the product is exposed to the clean room atmosphere have to be specially examined for contamination. Nowhere is this aspect more important than in the microbiological testing laboratory.

10.5.3 Commissioning of general pharmaceutical areas

Although special precautions have to be taken with products which have to satisfy sterility tests, the bulk of pharmaceutical products, which do not have to meet such stringent criteria, have more and more to meet tough purity conditions. Cross-contamination is a hazard in a multiproduct pharmaceutical factory and commissioning staff are very liable to carry traces of one product to another area on their persons, or more likely on their clothing, if they do not observe strict washing and changing requirements. The performance of the commissioning team in this respect will often set the standard for newly-recruited staff in a factory located in an area where pharmaceutical manufacturing behaviour is new. Commissioning staff who are not familiar with modern changing procedures, and the correct wearing of the garments used, should be sent to an established pharmaceutical factory for training. Faulty changing procedures result in endless commissioning runs before satisfactory product is made.

All equipment parts which come into contact with the product are inspected by the commissioning team to ensure cleanliness. Failure to do this, and taking any remedial cleaning action necessary, will result in more trial runs before satisfactory product results. It is true that the first run of product is a cleaning run but pharmaceutical products are generally very expensive, and even though rejected product can sometimes be reworked, thorough initial cleaning is cheaper than subsequent product and yield loss. The commissioning team must not accept the assurance of the precommissioning team that everything is adequately clean; it never is.

The commissioning team needs to pay attention to the testing of filters on the discharge to the atmosphere side of air handling and ventilation systems. Inadequacies in this area can result in cross-contamination of product areas and products, and in environmental hazards to the public.

10.5.4 Validation

Validation is a procedure designed to ensure that the equipment and procedures of a process will consistently yield a product of the required quality. This proving stage follows commissioning, but in the view of many is part of the commissioning process. Certainly, many health authorities will not allow a product to be sold in their countries unless the manufacturing facility has been validated.

Validation generally concentrates on those stages of the process which are vital for a quality product.

Terminal sterilization is such a stage. The validation team, which is often drawn from the future production unit, but can in a new unit be supplemented by selected staff from the commissioning team, will ensure the following:

• The trace or print-out showing the autoclave temperature, time and pressure cycle matches specification, and research data supports the chosen cycle. The instruments recording the data are properly tested, and testing is properly recorded. Signed certificates from the engineering department are necessary, as is the production of a planned maintenance procedure, duly authorized, to ensure as far as possible that verification will continue in the future.

• The steam used in the autoclave, if in contact with the product, is free of boiler additives and pipe scale. The team needs to see the written results of the steam analysis and the procedure for testing the final filter.

• The team needs to see trial evidence that the sterilization temperature recorded is present throughout the autoclave load. The results of regular monitoring of autoclave temperature across the whole load volume will have to be produced, to ensure there are no cold spots due to inadequate air removal or joint leakage. Samples of final product may have to be taken from various parts of the autoclave and tested for sterility, before the team can be satisfied. The number of samples taken will be far in excess of the number of normal batch samples.

Before validation begins, the engineering and analytical departments have to be fully operational, with written procedures for testing, and testing equipment proven. It can be argued that this should be the situation in any type of process plant, but it is often not the case; in the pharmaceutical plant it is essential.

Areas producing sterile products can take up to three months to validate, so validation is a procedure which must be planned into the overall project plan, and steps taken at the design and later stages to arrange matters allowing validation to be done as easily and quickly as possible.

10.5.5 The cost of commissioning pharmaceutical plant

When the Board of Directors authorizes a project, the decision to invest capital in the venture has been based on plant profitability. Real profitability depends on the return on the capital spent, when the return is made and how soon production of acceptable product is available at full flow sheet level.

When considering the cost of commissioning pharmaceutical plant, the validation stage must be included. The cost of commissioning is generally between 10% and 20% of the fixed capital project cost, and if the project cost is to represent, as it should, the real and complete cost of moving the project to the stage when acceptable product at full output level is achieved, then commissioning cost in its fullest sense must be included in the initial budget cost. The high cost of the loss of expensive pharmaceutical product, by way of total loss in the initial sub-standard batches plus the yield losses as the plant is worked up to full flow-sheet level, is included in the original submission to the Board.

Commissioning teams have a responsibility for a much greater share of the project cost and responsibility for providing full production on time than is generally imagined. Planning of the commissioning phases must be done in detail in the life of the project, as covered in the earlier chapters.

Problem avoidance in commissioning

11

Process plant projects vary greatly in their size, technology, the organizations and personalities involved and the problems encountered. The causes of these problems are less varied and can be remarkably similar on otherwise very different projects.

11.1 Typical problems

- lack of effective management of the project as a whole, because the project is not the responsibility of one person with authority over all the work;
- failures in planning and budgeting to allow for changes in the project 'environment';
- delegation of responsibility for commissioning to technical staff lacking the personality and experience to influence others not under their formal control;
- lack of defined objectives for commissioning;
- lack of attention to detail, particularly to the analysis of data obtained during commissioning;
- use of inappropriate managerial controls — for instance, applying rules for maintenance work to a major replacement of plant, or applying UK practice to work in other countries;
- reluctance to plan ahead, particularly to anticipate how to overcome faults, needs for additional manning, training and so on;
- failure to secure commitment to planning;
- inadequate specification of commissioning responsibilities and acceptance criteria in contracts;
- lack of control of subcontractors;
- failure to review the proposed designs of plant and process to ensure that they can be commissioned and operated safely and economically;
- failure to plan commissioning in detail to ensure that units, services and sub-systems can be checked and brought progressively into full operation;
- reluctance to consult specialists on problems and their remedies;

- design changes;
- mistakes arising from the acceptance of late changes to designs or to programmes;
- assumptions that licensed or patent designs will work without checking or testing;
- failures of housekeeping in not protecting and/or maintaining plant during manufacture, erection, testing or commissioning;
- dependence upon individuals' unrecorded knowledge of plant, services, operating principles and so on;
- reluctance to anticipate the anxieties or objections of people who appear to be affected by a project;
- lack of prior training of operators and supervisors.

11.2 Underlying causes of problems

- underestimating the complexity of completing a project. At the start of a project, engineers and managers tend to concentrate on the immediate commercial and technical decisions and make optimistic assumptions about the consequential effects;
- lack of confidence in planning, perhaps due to experience of previous failures to analyse problems and enforce decisions;
- slow uncoordinated responses to changes or problems affecting objectives after starting a project — for example, to changes in demand, problems of design, and cost suppliers;
- failure to recognize that projects affect the 'status-quo' and may require agreements to changes in jobs, relationships and procedures;
- inexperience in planning and controlling contracts;
- lack of a system for learning from experience, particularly a reluctance to analyse successful and unsuccessful decisions.

Glossary of terms

Analogue
Variable signal, proportional to process values.

Approved for construction
Engineering drawing and data which carry authorization for use in construction plant or equipment.

Arrow diagram
Planning diagram which shows by means of arrows the logical sequence of events and interdependence of activities

Bar-line
Programme in which activities are charted against time and durations shown by a bar, or line, drawn between scheduled start and finish dates.

Batch process
A process usually characterized by comparatively slow reactions with relatively large inventories of reactants and multistage reactions or process steps.

Benchmark
Common reference point.

Bias
Adjustment value to allow for error compensation.

Biomass
The living micro-organism 'product' formed in a biochemical reaction.

Blind
Blank plate inserted between flanges to effect positive isolation of lines, vessels or equipment.

Calibration
Setting up of instrument to match specific process conditions.

Card
Assembly of electronic components on a printed circuit board.

Chazop
Technique for identification of potential hazards and operability problems specifically associated with comuter control systems.

Clean room
Air conditioned environment with no radiation and an atmosphere essentially dust free.

Client
Organization placing contract for plant, equipment or services.

Coding
Entering software onto a computer.

Commissioning
(i) (specific term)
Putting into operation or otherwise making 'live', plant, equipment, systems or sections thereof, for normal duty. Adjusting and optimizing operating conditions for attainment of design performance and, where applicable, stipulated guarantees.
(ii) (collective term)
Unless otherwise defined, references such as 'commissioning', 'commissioning schedule', 'commissioning team' and so on, may be taken to include precommissioning and all related aspects of commissioning.

Commissioning manager/chief commissioning engineer
Person in overall charge of commissioning activities on a specific project.

Commissioning modifications engineer
Person responsible for monitoring and obtaining necessary authorization for all/any modifications to plant and equipment proved necessary during commis-

sioning operations and for ensuring records are properly amended in respect of same.

Completion of erection

Frequently, attainment of readiness for commissioning of plant, equipment, systems or sections thereof, in which case all pre-commissioning activities are included. In some instances, however, these terms may exclude precommissioning, thus denoting completion to the extent of stipulated mechanical tests having been satisfactorily fulfilled; precommissioning then follows.

Configuration chart

Definition of system software requirements.

Contactor

Motor control relays.

Continuous process

A process usually characterized by high throughputs and fast reactions with relatively low inventories of reactants under closely-controlled operating parameters.

Contractor

Organization carrying out specified works agreed under contract with client.

Control loop

An instrument loop that has a process control function.

Crafts personnel

Skilled workers specializing in specific technical disciplines.

Database

Base information.

Digital

Two-state presentation of values — for example, on/off.

Discrete faceplate

Single display showing loop information.

Earthing

Bonding of metalwork to low impedance source.

Emergency procedure
Action necessary to ensure safety of personnel and plant in event of sudden abnormal circumstances.

Fire permit
Certificate authorizing, under stated circumstances, use of naked flame in areas where ignition sources are normally prohibited.

Frame
Mechanical mounting arrangement.

Greenfield site
Proposed location of plant where no other installation already exists.

Grounding
Common reference point for electronic signals.

Hardware
Electronic components.

Hazop – hazard and operability studies
Technique for identification of potential hazards and operability problems.

Home office
Headquarters office responsible for project and which provides support for site teams.

IBC – intermediate bulk container
A proprietary portable container for the movement of chemicals/powders and so on (similar to 'Tote bin').

In-house
Work, often relating to a proprietary technology, carried out in owner's design offices.

Input/output (i/o)
Signals to and from a control system.

i/o hardware
Devices to which i/o are connected.

i/o Software
Logic which controls I/O.

Instrument loop
Related elements of a single measurement or control.

Interlock
Logic to prevent maloperation.

Licensor
Owner of proprietary system — for example, chemical process or software.

Line diagram
Detailing plant and utilities with lines and symbols from which engineering drawings are developed.

Loop diagram
Drawing showing all elements of an instrument loop.

Lump sum
Usually, selling price for technology, goods and/or services.

Main contractor
Organization having major responsibility for project under contract with client; often with responsibility for sub-contractors also.

Master line diagrams
Latest revision of line diagrams marked-up to include all changes and modifications.

Mechanical acceptance/mechanical completion
See 'completion of erection'.

Mimic diagram
A diagram which portrays plant, process and utility flows in miniature displays usually located in control room.

Milestones
Agreed objectives for completion of specific phases of the work.

Operating system
Standard software.

Operator entry code
See 'Password'.

Over-ranging
Driving an instrument outside its normal operating limits.

Owner
Proprietor; buyer.

Pacifying
The surface treatment of process pipework and vessels to reduce its surface activity and tendency to react with process chemicals.

Password
Keyboard sequence required before operator can control actions.

Pathogen
An organism or substance which causes disease.

Process and Instrument diagram (P&ID)
Line diagram showing the process plus the presence and function of instrumentation

Permit to work
Certificate authorizing specific work on plant or equipment in a normally operational area.

Precommissioning
Preparation, functional testing and making ready for commissioning plant, equipment, systems or sections thereof suitably completed and mechanically tested — for example, nitrogen purging; line-vessel cleaning.

Precedence diagram
Component of planning method.

Peripheral
Auxiliary device connected to a computer — for example, printer's disk drives).

PLC
Programmable logic controller

Process commissioning
See 'Commissioning'.

Purging
Displacing potentially hazardous gases or vapour from plant or equipment with inert gas, or sweetening by displacing inert gas with air.

Pyrogen
An agent which if taken into the body produces fever systems.

Quality assurance
System which ensures 'Fitness for Purpose' at all stages of project.

Rack
Mounting arrangement for printed circuit boards.

Range
Minimum and maximum values.

Reimbursable contract
A contract where payment for services performed is at agreed rates, and materials and so on supplied are charged for on an agreed basis.
NB: 'stage payments' may be made at defined intervals of work completion.

Relay
Electro-mechanical switching device.

Reservation list
Items against which a qualified acceptance of plant or equipment is given.

Retrofit
Modifications of existing plant usually to an updating of design.

Rheology
The detailed properties of a fluid in motion usually expressed as the relationship between shear stress and rate of shear.

Sensor
Measuring device.

Sequence
Succession of logical operations.

Sequence flow digram
Representation of the interaction of sequences.

Service system
A system providing an essential service to the process — for example, cooling water, power and steam.

Signal conditioning block
Conversion of a signal from one type to another — for example, analogue to digital).

Site
Location where plant is to be installed.

Slip plate
Similar to 'blind'.

Software
Coding written to instruct a computer system

Applications software
Process plant specific software.

Configurable software
Software that can be changed easily by plant operators.

Systems software
Software that is operating systems specific.

Spade
Similar to 'blind'.

Systems diagnostics
Facility for fault finding provided by computer.

Tag number
Unique identification of an instrument loop element.

Testing

Verifying performance by measurement or operation.

Tote bin

A proprietary standard portable container for the movement of chemicals/powders, and so on (similar to 'IBC').

Transducer

Sensing element.

Turnkey contract

A lump-sum price contract covering the entire scope of work, usually from conceptual design to acceptance of the work, including direction of commissioning. NB: "Stage payments" may be made at defined intervals of work completion.

Utilities

Services to process plant area such as power, water, air, steam, inert-gas; sometimes referred to as "off-sites".

Validation

Checking.

VDU

Visual display unit.

Vendor

Seller.

Venting

Permitting discharge to, or ingress from, atmosphere.

Vessel entry

Usually a certificate permitting entry to a vessel which might be hazardous in normal use.

Watchdog

Automatic check of correct operation of a programmable control system.

Works

Factory.

References

1. IChemE, 1995, *Model Form of Conditions of Contract for Process Plant Suitable for Lump Sum Contracts in the United Kingdom* (the Red Book), 3rd edition.
2. IChemE, 1992, *Model Form of Conditions of Contract for Process Plants Suitable for Reimbursable Contracts in the United Kingdom* (the Green Book), 2nd edition.
3. IChemE, 1997, *Model form of Conditions of Contract for Process Plant Suitable for Subcontracts in the United Kingdom* (the Yellow Book), 2nd edition.
4. *BS EN ISO 9001, Quality Systems: specification for design, development, production, installation and servicing*, 1994.
    BS EN ISO 9002, Quality Systems: specification for production, installation and servicing, 1994.
    BS EN ISO 9003, Quality Systems: specification for final inspection and test, 1994. (British Standards Institution, UK).
5. Ashley, R., *The installation and commissioning of process plants with fully integrated control systems,* IMechE Conference, London (November 1984).
6. *BS 6739, Care of practice for instrumentation in process control systems: installation design and practice*, 1986 (British Standards Institution, UK).
7. Health and Safety Executive, 1987, *Programmable Electronic Systems in Safety Related Applications, Parts 1 and 2* (The Stationery Office (formerly HMSO, UK).
8. Institution of Electrical Engineers, 1985, *Guidelines for the Documentation of Software in Industrial Computer Systems.*
10. Briggs, M., Buck, S. and Smith, M., 1997, *Decommissioning, Mothballing and Revamping* (IChemE, UK).
11. IChemE, 1981, *Model Form of Conditions of Contract for Process Plant Suitable for Lump Sum Contracts in the United Kingdom* (the Red Book), 1st edition.

Appendices

Note: Appendices are numbered according to the chapter in which they are referenced.

Appendix 2.1

Defining work and responsibility of construction and commissioning by contractor and the client

Phase A – Construction and precommissioning

Prepare plant/equipment for precommissioning/mechanical testing

Legend: C = Client   H = Contractor   V = Vendor

Work items		Supervised by	Trades & labour by	Inspected by
1.	Ensure adequate safety precautions and services available for activities in this and ensuring phases	Phase A–D H Phase E–G C	H C	C & H C & H
2.	Install service gland packing in all minor machinery and drivers	H	H	H
3.	Install service gland packing and lubricate all types of valves	H	H	H
4.	Check alignment and lubrication of all minor rotating machinery and drivers	H	H	H
5.	Fill electrical equipment with oil as required	H	H	H
6.	Block-off or isolate all equipment before line flushing and testing	H	H	H
7.	Remove bellows-type expansion joints and rupture discs from pipework, etc before line flushing and testing	H	H	H
8.	Remove installed relief valves and any instruments liable to damage from lines and vessels if assembly has been made for piping fit-up checks, etc before line flushing and testing	H	H	H
9.	Correct any construction errors or omissions	H	H	H

Note: an alternative form of check-list can be found in the American Petroleum Institute (API) publication API 700.

Appendix 2.2

Defining work and responsibility of construction and commissioning by contractor and the client

Phase B – Construction and precommissioning

Prepare services; clean and pressure test systems

Legend: C = Client H = Contractor V = Vendor

	Work items	Supervised by	Trades & labour by	Inspected by
1.	Precommission/prepare for duty utility systems and services, including package items, as required for testing/precommissioning plant systems and equipment	H	H	C & H
2.	Check electrical installations for power, lighting, and instrumentation for operability and safety	H	H	C & H
3.	Check for correct rotation of minor rotating machinery drivers and carry out uncoupled run	H	H	H
4.	Hydraulic and/or pneumatic pressure test equipment to specification requirements to check connections and joints for pressure tightness	H	H	C & H
5.	Clean all lines of loose material by flushing, steaming or blowing	H	H	H
6.	Hydraulic and/or pneumatic pressure test lines to specification requirements	H	H	C & H
7.	Apply any special preparation on inside surfaces of lines as specified	H	H	H
8.	Set up relief valves on test rig for witness testing and subsequently reinstall in lines	H	H	C & H
9.	Install orifice plates after line cleaning and testing	H	H	H
10.	Remove all loose material and dirt from inside vessels, columns and tanks; check/install internals and packings and 'close-up' upon approval of cleanliness and inspection of internals	H	H	C & H
11.	After completion of line/equipment pressure testing, remove all swing blinds (slip-plates) from lines other than those required for operation purposes and install temporary strainers	H	H	H
12.	Correction of construction errors and omissions	H	H	H
13.	Retest after corrective work or alteration	H	H	C & H

Appendix 2.3

Defining work and responsibility of construction and commissioning by contractor and the client

Phase C – Construction and precommissioning

Check and prepare major mechanical equipment, instrumentation, and protection systems

Legend: C = Client   H = Contractor   V = Vendor

Work items	Supervised by	Trades & labour by	Inspected by
1. Carry out any chemical cleaning required — for example, boiler system, compressor suction pipework and lubricating systems and so on	H & V	H	C H V
2. Check alignment and lubrication of all major drivers, compressors and blowers which have been assembled at site	H & V	H	C H V
3. Clean and flush lubricating oil and seal oil installations and circulate oil in all main drivers, compressors and blowers	H & V	H	H
4. Check for correct rotation of major drivers and carry out uncoupled run	H & V	H	C H V
5. Carry out short running-in tests with air on reciprocating compressors to check bearings, rods, governors, safety devices and ancillary equipment	H & V	H	C H V
6. Carry out short running-in tests with air (where suitable) on all main centrifugal blowers and compressors to ensure satisfactory mechanical operation when a works test has not been carried out	H & V	H	C H V
7. Carry out short running-in tests on pumps with water (where suitable) to ensure satisfactory operation	H & V	H	C H V
8. Check action of instruments, continuity of thermocouple connections and instrument circuits. Check control valves are correct for direction of flow and action on air failure	H & V	H	C H V
9. Test instruments and loop-check control circuits	H & V	H	C H V
10. Test electrical controls and plant protection systems	H	H	C & H

Appendix 2.4

Defining work and responsibility of construction and commissioning by contractor and the client

Phase D – Construction and precommissioning

Final preparation for start-up

Legend: C = Client   H = Contractor   V = Vendor

Work items	Supervised by	Trades & labour by	Inspected by
1. Check provisions for fire-fighting, safety equipment and communications	H	C	C & H
2. Check that utility systems are fully precommisioned	H	C	C & H
3. Calibrate instruments	H & V	H	C & H
4. Check that orifice plates and permanent blinds are installed	H	H	H
5. Charge catalyst(s) in accordance with recommended procedures	H comm	C	C & H
6. Carry out any chemical pretreatment of process vessels required — for example, gas absorption plant vanadation	H comm	H	C & H
7. Dry out refractory linings other than when included in plant start-up procedures	H	C	H
8. Dry out and purge lines and equipment	H	C	H
9. Test for leak-tightness of plant/systems (standing test)	H	H	C & H
10. Remove, clean and replace temporary strainers	H	C	H
11. Check drainage systems clear and liquid seals filled	H	H	H
12. Check completeness of battery limit connections	H	C	C & H
13. Check battery limits disposal systems are complete and ready for function	C	C	C
14. Check relief systems to flare or blow-down to see no blockages	H	H	H
15. Adjust pipe supports for expansion and loading strains	H	H	H
16. Correct construction errors and repairs	H	H	C & H

Appendix 2.5

Defining work and responsibility of construction and commissioning by contractor and the client

Phase E – Commissioning

Charge with feedstock and so on, start-up plant and operate

Legend: C = Client   H = Contractor   V = Vendor

	Work items	Supervised by	Trades & labour by
1.	Put utility systems into service in readiness for plant start-up, ensuring that any necessary chemical dosing procedure is established	H	C
2.	Charge feedstock/fuel and other systems fed from outside battery limits, consistent with appropriate safety precautions	H	C
3.	Commence start-up procedures including any necessary refractory dry-out, catalyst conditioning and so on, and proceed to bring plant on-line in accordance with operating manual and any supplementary instructions	H	C
4.	Check supports of all hot and cold lines and adjust for expansion strains	H	C
5.	Tighten hot joints	H	C
6.	Optimize operating adjustments; agree procedures for plant performance test(s)	H	C
7.	Maintenance, routine cleaning and normal adjustments to plant	C	C
8.	Complete lagging and painting	H	H
9.	Complete minor construction details	H	H
10.	Record any modifications carried out during commissioning and annotate drawings/diagrams accordingly	H	—
11.	Clean up site	H	H

Appendix 3.1

Reservation check-list

Pipelines and pipework

When carrying out plant reservations checks, the following list of possible faults should be looked for:

		Tick when checked
1.	Screwed plugs in pipes, only permissible on air, water, nitrogen under 100 psi, one and a half inch nominal bore and below	
2.	Faulty welding	
3.	Correct joints	
4.	Odd sized bolts	
5.	Black bolts in cold joints	
6.	Faulty pipeline supports	
7.	Pipe not resting on supports	
8.	Are expansion slippers safe? Will they push off structure when line is hot?	
9.	Check spring hanger settings	
10.	Faulty spring hangers	
11.	Low point drains fitted where necessary, and high point vents	
12.	Lagging — missing, damaged, loose and so on	
13.	Vent and drain blanks fitted where necessary	
14.	'Weep holes' in relief valve exhaust lines, only on non-hydrocarbon or atmosphere RVs	
15.	Have all slip plates been removed and spec plates turned?	
16.	Spring hangers — have restraining pins been removed?	
17.	Make sure pipework is up to the P&I diagram specification	
18.	Necking off hazard — is there any equipment, or small bore pipe projection that can be accidentally broken off?	
19.	Are small branches — that is, drains — sufficiently clear of pipe supports?	
20.	Do drain lines run to underground drains? They should not flow over paved areas	
21.	Flanges lagged up	

Appendix 3.2

'Chemicals and hydrocarbons in' check-list

Jobs to be done before chemicals or hydrocarbons can be safely brought into the plant	Action	Sign	Date
Complete all necessary reservations			
Nitrogen purge systems prepared and leak tested			
Blowdown system 'live'			
Oil/water separator in commission			
New perimeter fence erected, with 'Dematched Area' notices			
Dematching hut in position and gateman available			
Brief all construction personnel on consequence of plant becoming a 'Dematched Area'			
Fire alarm I/C and all positions tested			
Compressed air sets in position			
Fire extinguishers in position			
Eye wash bottles in position			
Personnel showers in position and checked			
Fire hoses in position			
Fire main I/C and check that hoses fit hydrants			
Drench water sprays tested			
Fire steam hoses in position and check they connect securely			
Fire permits in use			
Flush drains to prove them free from obstructions			
Plant areas clean (fire hazards removed, such as rags)			
Check plant lighting			
Remove contractor's buildings, tarpaulins and so on			
Remove non-flameproof equipment			
Nominate shift fire teams and arrange practice alarm			
List all possible welding jobs — get most done beforehand			
Check that welding sets pass works electrical inspection			
Check gas detectors			
Obtain 'Means of Escape' certificate			
Check that segregation plates can be easily removed			
Inform fire station			
Invite safety department to inspect and pass comment			
Carry out thorough search for matches			
Make sure that neighbouring plants know how this affects them			
Inform services (affect their drains)			
Inform Factory Inspector and Alkali Inspector			
Inform local authorities			
Inform records section			

Final check by plant managers ....

Date   ....

Time   ....

Appendix 3.3

Safety assessment

Plant: Title: Reg number:

Underline those factors which have been changed by the proposal.

Engineering hardware and design
line diagram
wiring diagram
plant layout
design pressure
design temperature
materials of construction
loads on, or strength of:
    foundations
    structures
    vessels
    pipework/supports/bellows
temporary or permanent:
    pipework/supports/bellows
    valves
    slip-plates
    restriction plates
    filters
instrumentation and control systems
trips and alarms
static electricity
lightning protection
radioactivity
rate of corrosion
rate of erosion
isolation for maintenance
    mechanical-electrical
fire protection of cables
handrails
ladders
platforms
walkways
tripping hazard
access for
    operation
    maintenance
    vehicles
    plant
    fire-fighting
underground/overhead:
    services
    equipment

Process conditions
temperature
pressure
flow
level
composition
toxicity
flash point
reaction conditions

Operating methods
start-up
routine operation
shutdown
abnormal operation
preparation for maintenance
emergency operation
layout and positioning of controls
    and instruments

Engineering methods
trip and alarm testing
maintenance procedures
inspection
portable equipment

Safety equipment
fire-fighting and detection systems
means of escape
safety equipment for presonnel

Environmental conditions
liquid effluent
solid effluent
gaseous effluent
noise

Appendix 3.3 (continued)

Within the categories listed below, does the proposal:	Yes or no	Problems and actions recommended	Signed and date
Relief and blowdown			
1. Introduce or alter any potential cause of over/under pressuring (or raising or lowering the temperature in) the system or part of it?			
2. Introduce a risk of creating a vacuum in the system or part of it?			
3. In any way affect equipment already installed for the purpose of preventing or minimizing over or under pressure?			
Area classification			
4. Introduce or alter the location of potential leaks of flammable material?			
5. Alter the chemical composition or the physical properties of the process material?			
6. Introduce new or alter existing electrical equipment?			
Safety equipment			
7. Require the provision of additional safety equipment?			
8. Affect existing safety equipment?			
Operation and design			
9. Introduce new or alter existing hardware?			
10. Require consideration of the relevant codes of practice and specifications?			
11. Affect the process or equipment upstream or downstream of the change?			
12. Affect safe access for personnel and equipment, safe places of work and safe layout?			
13. Require revision of equipment inspection frequencies?			
14. Affect any existing trip or alarm system or require additional trip or alarm protection?			
15. Affect any reaction stability or controllability of the process?			
16. Affect existing operating or maintenance procedures or require new procedures?			
17. Alter the composition of, or means of disposal of effluent?			
18. Alter noise level?			

Safety assessor .... Date ....

Checked by .... Plant manager   Checked by .... Engineer

Appendix 5.1

Typical check-list

The IChemE Red Book (see reference 11) gives a useful *aide-mémoire* for mechanical comissioning which is reproduced below:

1. Installation of gland packing and lubrication of valves and minor machinery, checking rotation of drivers and so on.

2. Isolating equipment, removing expansion joints, rupture discs and relief valves from pipework and equipment before line flushing, cleaning and testing.

3. Cleaning of pipework by flushing, steaming, blowing and so on.

4. Hydraulic and/or pneumatic pressure testing of equipment and pipework.

5. Application of special treatment or other preparation of inside surfaces.

6. Testing of relief valves.

7. Removing loose material and dirt, installation of internals, inspection and closing up of vessels and so on.

8. Installation of orifice plates after line cleaning, installation of temporary strainers, removal of slip-plates and so on.

9. Checking electrical installations.

10. Checking alignment of major machinery and drivers' cleaning and flushing lubricating oil installations and so on.

11. Carrying out short running-in tests on machinery as required.

12. Checking action of instruments and control valves, testing electrical controls and alarms and so on.

13. Calibration of instruments.

14. Drying out refractories.

15. Adjustment of pipe supports for expansion.

16. Removal, cleaning and replacement of temporary strainers.

17. Checking orifice plates and permanent blanks, relief systems, safety provisions and so on.

18. Charging catalysts as required.

19. Charging raw materials, process chemicals, fuel and so on, as required.

20. Warming-up, starting-up fluid flows and so on, as required.

21. Tightening of hot joints.

22. Starting-up and operating the various plant items.

23. Routine maintenance, cleaning, plant adjustments and so on.

Appendix 5.2.1

Pressure vessels: inspection

Orientation		
Alignment and level	Vertical Horizontal	
Method of fixing	Holding down bolts Grouting Davit Agitator motor/gearbox/support	
Nozzles Manway Handhole Vent Overflow Drain		
Pressure relief	Valve Bursting disc Setting labels correct	
Shell mounted instruments fitted per P&ID or ELD	Temperature Flow Pressure Level	
Insulated Fire-proofed Painted Vessel identification added		
Electrical lighting	Permanent Emergency Level indicators Viewing	
Internal visual inspection		
Installed and fixed	Baffles Overflows Weirs Downcomers Trays Chimneys Internal distribution systems Dip pipes Vortex breakers Packaging support grids Hold down grids Packing Demister Agitator	

Alignment to BS 5276 Part 3 or equivalent and settings to instructions	Tray level tolerance
	Weir set height
	    tolerance
	Downcomer outlet set height
	    tolerance
	Distribution system tolerance
		±
		
		±
		
		±
		±

Appendix 5.2.2

Equipment check-out schedule — packed columns Sheet 1 of 2	Job number: Client: Site:				
System/code number:	Vessel number:				
Process data sheet:	Design data sheet:				
Equipment:	Tag number:				
Carried out at:	Engineering line diagram number:				
Check-list items	Initialled as witnessed:				
	Date	Contractor	#	Client	Certification authority

Gas inlet/reboiler vapour return deflector(s) fitted and secured as in drawing
Weir in base of vessel located as in plant design schematic (PDS) with relation to nozzles
Liquid baffle(s) vortex breaker fitted
Gas inlet pipe installed and flow area dimension and location as on PDS
Demister pad(s) in gas/vapour outlet(s) properly secured
Liquid inlet distributor installed properly (including spray direction) and with flow area dimensions as in drawing
Liquid distributor tray(s) installed
Wash trays installed, sealed and secured correctly — satisfactory leakage rate test carried out
    (trays number ....)
Gas redistribution chimney(s) installed properly with flow area dimensions as in drawing
Draw-off trays — installed, sealed and secured correctly — satisfactory leakage rate test carried out
    (trays number ....)
Dimensions and number of holes/bubble-caps on wash tray(s) as on PDS
Weir tray heights as in process data sheet
Downcomers — width and clearance at bottom lip as in process data sheet					

References	# = E — Construction O — Commissioning I — Inspection (QC)				

The above check-out was completed to our satisfaction:

For contractor: **For client:**   **For certification authority:**

Signature   Date   Signature   Date   Signature   Date

95

Appendix 5.2.2 (continued)

Equipment check-out schedule — packed columns Sheet 2 of 2	Job number: Client: Site:				
System/code number:	Vessel number:				
Process data sheet:	Design data sheet:				
Equipment:	Tag number:				
Carried out at:	Engineering line diagram number:				
Check-list items	Initialled as witnessed:				
	Date	Contractor	#	Client	Certification authority
Drain fitted, as in drawing Manway in vessel bottom weir installed, secured and sealed correctly Baffles correctly positioned in column top, bottom and draw-off section Vessel lining as specified in drawing Column interior, internals and nozzles clean prior to loading packing Packing support grid(s) installed properly and same dimensions as in drawing Packing discharge spider fitted Packing separated from fines/degreased/pretreated and protected Packing installed, as detailed below, hold-down grid(s) installed on bed(s) Instruments (including pressure/analysis points) correctly located to fulfil requirements of process data sheet Check made to ensure proper final joints fitted in vessel flanges Flushing and chemical pretreatment carried out					
References	# = E — Construction O — Commissioning I — Inspection (QC)				

The above check-out was completed to our satisfaction:

For contractor: **For client:**   **For certification authority:**

Signature   Date   Signature   Date   Signature   Date

Appendix 5.3.1

Equipment check-out schedule — control panels	Job number: Client: Site:

System/code number:	Vessel number:
Process data sheet:	Design data sheet:
Equipment:	Tag number:
Carried out at:	Engineering line diagram number:

Check-list items	Initialled as witnessed:				
	Date	Contractor	#	Client	Certification authority

Correct materials used, finishing as per
  specification
Connecting hardware provided as per design
Piping and wiring devices provided as per design
Piping at rear of panel secure and of neat
  appearance
Wiring at rear of panel secure and of neat
  appearance
Tubing and wiring identified by markers
Annunciator panels comply with design
Air filters installed and readily accessible
Instruments located as per design
Instruments correctly labelled front and rear
Circuits checked for continuity
Circuits checked for accuracy of connection
System isolation tests complete (using 250V DC)
Pneumatic tubes/connections checked for leaks
Functional test via simulated field input and output
Chart drives checked where applicable
Automatic/manual selection proven where
  applicable
Fit of doors/catches/slides checked
Free of any transit damage					

References	# = E — Construction
	     O — Commissioning
	     I — Inspection (QC)				

The above check-out was completed to our satisfaction:

For contractor: **For client:**   **For certification authority:**

Signature   Date   Signature   Date   Signature   Date

Appendix 5.3.2

Equipment check-out schedule — orifice plates	Job number: Client: Site:

System/code number:	Vessel number:
Process data sheet:	Design data sheet:
Equipment:	Tag number:
Carried out at:	Engineering line diagram number:

Check-list items	Initialled as witnessed:				
	Date	Contractor	#	Client	Certification authority

Tag number	Orifice bore		Orifice	Installed	Date	Contractor	#	Client	Certification authority
	Design	Ovality	Finish*	Correct*					
		A							
		B							
		A							
		B							
		A							
		B							
		A							
		B							
		A							
		B							
		A							
		B							
		A							
		B							

* Tick when checked

References	# = E — Construction O — Commissioning I — Inspection (QC)

The above check-out was completed to our satisfaction:

For contractor: **For client:** **For certification authority:**

Signature Date Signature Date Signature Date

Appendix 5.3.3

Equipment check-out schedule — relief valves	Job number: Client: Site:

System/code number:	Vessel number:
Process data sheet:	Design data sheet:
Equipment:	Tag number:
Carried out at:	Engineering line diagram number:

Check-list items						Initialled as witnessed:				
						Date	Contractor	#	Client	Certification authority
Tag no RV	Design set (hot)	Design set (cold)	Actual set	Leak rate	Blow down ring reset	Exh'st direct-ion safe				

Note: pressure units used

References	# = E — Construction O — Commissioning I — Inspection (QC)

The above check-out was completed to our satisfaction:

For contractor: **For client:**   **For certification authority:**

Signature   Date   Signature   Date   Signature   Date

Appendix 5.3.4

Piping systems test-sheet

Design and test data		
Availability of the following data relevant to the piping system	Engineering line diagram Piping test schedule Piping arrangement drawing Pipeline schedule Piping and valve specifications Piping isometrics Piping test procedure Piping fabrication and erection codes Pipe support details and schedule	
Test certificates for	Off-site fabricated piping Heat treatment or other Process/utility valves, control valves etc NDT (X or Gamma Ray and so on) In-line process equipment	
Piping and valve material analysis certificates		
Welders' qualification certificates		
Pretest visual inspection		
Routing and size correct to ELD/arrangement drawing/piping isometric		
Installation of piping and piping components complete	Joints, bolts, nuts and gaskets, expansion loops and bellows	
	Fixed anchors, sliding supports, guides spring or fixed hangers	
	Jackets/jumpers, tracing, conductive bolts/earthing straps	
Installation/orientation with respect to flow	Process and utility valves, non-return and relief valves, process control valves, orifice plates and flowmeters	
Location of in-line components for access, operation, maintenance and safety of operatives		
Location and installation of	Vents, drains, drip legs, drip rings, utility station connections, steam traps, filters, strainers line blinds, spectacle plates, by-passes, instrument tapping points for pressure, temperature and flow, plugs and rodding out points	
Installation of field mounted instruments such as thermowells and pressure gauges		
Check and ensure system is devoid of insulation and paint		

Appendix 5.3.4 (continued)

Pretest preparation	
Prepare a written test plan and mark engineering line diagram (ELD)/general arrangement (GA) with location of all spools, spades, blanks, vent valves, strainers and so on, agree with site manager and initiate	
Obtain spades, blanks, strainers, vent valves, bolts, nuts, gaskets and fabricate pipe spools	
Remove relief valves (RVs) for bench testing and orifice plates for checking, make good joints or blank off	
Remove or spade off any control valve or instrument liable to damage under test pressure, replace with spool or make good joint and open any by-pass valves	
Spade off or isolate process equipment with lower allowable pressure than test pressure	
Spade off all overflows, close drains, fit vent valves and ensure test medium available	
Calibrate test gauge and check range adequate for test pressure and detect pressure loss	
Define testing fluid	

Pressure test procedure		
Site manager's clearance obtained, other contractors informed and safety notices positioned		
Ensure all test personnel are competent and briefed regarding extent, duration and limits of test		
Open up system flush or blow through to remove mill scale/rubbish, fit temporary strainers and close		
Hook up test pump to line and test medium, open vents and commence filling system		
Conduct test per test procedure, attend to remedial works and bring to test pressure and hold		
Invite client's representative to witness test		
Prepare test certificate and record	System title, line numbers and plant reference Numbers, date, time and duration of test Pressure, test certificate number and obtain client's signature	

Post-test procedure	
Open up, drain down, remove and account for all spades, blanks, spools, plugs, vent valves, strainers and test equipment, flush or blow through and dry out	
With new gaskets, reinstall all bench tested RVs, orifice plates, control valves, thermowells, flowmeters, pressure gauges and remove safety notices and so on	
Check installation complete and purge or chemically clean if part of take-over procedure	
Complete construction works — for example, paint, insulate, colour code and so on	

Appendix 5.4.1

Typical electrical system check-list

1. Examine transformers for mechanical damage, oil leaks and so on. Check no-load tap-changer for proper movement.

2. Inspect switchgear for damage and missing parts, alignment, clearance of moving parts and so on.

3. Verify that instrument transformers, instruments, relays, fuses and other devices are of proper type, size and rating.

4. Test or check direct trip breakers.

5. Perform insulation resistance test (when appropriate).

6. With breaker in test position, test operation with local, remote and manual operation.

7. Test automatic transfer equipment by simulating power failure and under-voltage conditions. Set all relays in accordance with job relay schedule and check operation.

8. Check fuse holders for damage and fuses for size and rating. Witness relay tests and settings.

9. Test cables rated at less than 5000 volts with 500 volt-megohm instrument.

10. Check rotation and ability of motor to synchronize.

11. With motor running, check bearing and winding temperature, rotation and vibration.

12. Check battery and charger for damage and check electrolyte for level and specific gravity. Adjust charging current and voltage.

13. Check emergency generator and automatic starting generation and transfer with simulated loss of normal power.

Appendix 5.4.2

Equipment check-out schedule — **general electrical installation**	**Job number:** **Client:** **Site:**				
System/code number:	Vessel number:				
Process data sheet:	Design data sheet:				
Equipment:	Tag number:				
Carried out at:	Engineering line diagram number:				
Check-list items	Initialled as witnessed:				
	Date	Contractor	#	Client	Certification authority

Cable tray routes as per drawing
Cable tray supports adequate and secure
Cable tray assembled and secured correctly
Cable tray earthing as per specification
Cable tray earth continuity checks satisfactory
Earth cables/bonds securely terminated
Cables securely banded/shrouded
Correct cable glands fitted
Junction boxes and covers securely fitted
Flameproof equipment fitted as per specification
Lighting fittings correctly erected and secure
Distribution boards installed and connected as
  per drawings
Local control stations installed and connected as
  per drawings
All equipment clearly labelled
Cables and terminations clearly identified
Area and equipment clear of debris/surplus
  material					

References	# = E — Construction
	     O — Commissioning
	     I — Inspection (QC)				

The above check-out was completed to our satisfaction:

For contractor: **For client:**   **For certification authority:**

Signature   Date   Signature   Date   Signature   Date

Appendix 6.1

Loop check-list

Input channel
Check channel number
Correct tag number?
Correct scaling: range/bias/units?
Test input signal range: top/middle/bottom
Is signal characterization necessary — for example, square root and linearization?
Does the input signal need filtering?
If so, check the filter constant is sensible
Check alarm limits: hihi/hi/lo/lolo
Check sampling frequency

Faceplate/bar display
In correct overview/group?
Right type of template used?
Correct tag number?
Check information as per input channel
Correct engineering units shown?
Is set point shown OK?
Is fail safe position of output shown correctly?

Trend display
Correct tag number?
Is the time-scale appropriate?
Check display frequency
Are range/bias/units of vertical scale OK?
Correct colour coding?
Does output signal need to be trended?

Archiving
Correct tag number?
Correct logging frequency?
How long is data to be archived for?
Check scope/need for data compression
Are engineering units OK?

Alarm handling
Is alarm put into alarm list directly?
Is alarm in right grouping?
Check priority/colour coding is appropriate
Is alarm logged on printer with date/time stamp?
Do tag number and alarm description correspond?
Can operator override alarm?
Any special requirements for annunciation?
Are standard facilities for acknowledgement OK?
Check any linked trips or interlocks

Appendix 6.1 (continued)

Control functions

Is set point local or remote? If local, is its value set correctly or if remote, is its
   address correct?

Check set point ramp rate

Is set point tracking required?

Does access to loop status functions need to be inhibited?

Can operator change set point in AUTO?

Can operator change output in MANUAL?

Does loop need to be disabled during start-up?

Are hi/lo alarms on error signal OK?

Has correct control algorithm been selected?

Is proportional action OK regarding oscillation?

Is integral action OK regarding offset?

Is derivative action OK regarding speed of response?

Does loop satisfy other performance criteria — for example, stability and noise?

Is the controller action in the correct direction?

Is the controller output bias sensible?

Are there any alarms on the output signal?

Output channel

Check channel number

Correct tag number?

Correct scaling: range/bias/units?

Are output channel and controller output consistent?

Check valve opening at maximum and minimum output signals

Check that valve action is fail safe

Is there any constraint on the output signal — for example, upper limit on valve opening?

Check functioning of any limit switches

Are time delays for testing change of status sensible?

Check output signal sampling frequency

Documentation

Are the loop configuration and P&I diagram consistent?

Is the loop diagram correct?

Are the configuration tables correct?

Is the loop tag number correct throughout?

Is the operator's manual complete?

Is the database listing up-to-date?

Have the disks been updated?

Appendix 6.2

Sequence check-list

Declarations
Is sequence number/name correct?
Check variables correctly named: integer/flag/block/signal/floating point and so on
Are variable types correctly defined?
Have constant values been correctly assigned?

Structure
Are the sequence steps in the right order?
Are the criteria for progression from step to step OK?
Are there any 'neverending' waits/loops?
Check logical for all branching and confirm no 'loose ends'
Are correct subsequences called?
Have subsequences got correct arguments?

Timing
Are all timing constants sensible/correct? Check both absolute and lapsed times
Are there any unnecessary waits/delays?
Have flight times been allowed for?
Sufficient time for discrepancy checking?
Check synchronization with other sequences

Contention handling
Have criteria for handling contentions been specified? First come first served basis?
Are criteria same throughout the sequence?
Check logic for handling contentions
   Are all relevant signals considered?
   Are correct flags set/reset?

Operator access
Is manual intervention required?
Does operator have full control over sequence? Check start/stop/hold/restart
Can sequence be stopped anywhere?
Can the sequence be progressed manually?
Does it have to be restarted from the same step? Check on need to skip/repeat steps
Does sequence need to override operator actions?

Recovery options
Are correct criteria used for initiating recovery?
Are criteria same throughout sequence?
Check on reserved/released variables
Are recovery actions correct — for example, branch forward/back, hold and shutdown?
Are actions same throughout sequence?
Branch into separate sequence/subsequence? If so, branch back to same place?

Appendix 6.2 (continued)

Recipe handling

Does the operator assign the recipe to the sequence?
Are the quantities of reagents correct?
Can the operator change the quantities/formulation?
What provision is there for rejecting out of range values?
How are 'as weighed' values handled?
Are the reagents listed in the correct order?
Check the channel number for each reagent feed tank
Are the operating conditions specified in the recipe OK?
    Check setpoints, ramp rates, time delays and so on
Are the logical values initialized for the start of a batch?
    Check discrete outputs, flags and integer variables
Have variable alarm limits been defined?

Firmware interface

Is loop status changed from within sequence?
    Confirm output OK is put into MAN
    Confirm setpoint OK if put into AUTO
Is configuration changed from within sequence?
    Check correct blocks/pointers used. Check correct parameters set up
    Confirm alarm settings OK
Needs to activate/suspend other sequences?

Sequence display

Is sequence in correct display group?
Is sequence number/name correct?
Confirm messages fit into reserved area
Check messages appropriate to steps
Are operator prompts intelligible?
Is annunciation/acknowledgement of prompts OK?

Batch logging

Is standard batch log in use? Check format appropriate
What identification is required? Batch/lot number/sequence recipe
Is all recipe information included? What about 'as weighed' values?
Are all actions/events logged? If not, check that appropriate ones are
Check provision for logging abnormal conditions
    Recovery options/manual intervention
    Maximum/minimum values of signals

Note: for sequence terminology, refer to reference 6.

Appendix 6.3.1

Instrument loop check-sheet

(Appendix E of BS 6739, 1986, reproduced by permission of BSI)

Instrument loop check-sheet			
Client: Client's project number:		Plant: Project Number:	
Loop number Line or equipment number		Service Pipe number	
Mechanical checks/electrical checks			
Measuring element:	Installation correct Isolating valves correct Tapping(s) position correct	☐ Location correct ☐ Materials correct ☐ Orifice diameter _____	☐ ☐
Impulse connections:	Correct to hook-up Pressure tested Steam/electrically traced	☐ Materials correct ☐ Test pressure_____ ☐ Lagged	☐ ☐
Field instrument(s):	Installation correct Weather protected	☐ Air supply correct ☐ Power supply correct	☐ ☐
Panel instrument(s)	Installation correct Scale/chart correct	☐ Air supply correct ☐ Power supply correct	☐ ☐
Control valves:	Installation and location correct Stroke tested Limit switch(es) set	☐ Size and type correct ☐ Positioner checked ☐ I/P transducer checked	☐ ☐ ☐
Air supplies:	Connections correct to drawings ☐	Blown clear and leak tested	☐
Transmission-pneumatic:	Lines inspected, blown clear and leak tested		☐
-electrical:	Insulation checked core-to-core Continuity checked Earth bonding checked	☐ Core to earth ☐ Loop impedance checked ☐ Zener barriers correct	☐ ☐ ☐
Temperature loops:	T/C or R/B checked Continuity checked	☐ Cable to specification ☐ Loop impedance checked	☐ ☐
General:	Supports correct	☐ Tagging correct	☐
Checked by:	Date:	Witnessed by:	Date:

Appendix 6.3.1 (continued)

Loop test:

	Transmitter input	Transmitter output	Local instrument reading	Panel instrument reading	
Measurement					

	Controller ouput	Transducer output	Valve positioner output	Control valve position	
Control					

Remarks:

Checked by: Date:	Witnessed by: Date:
Accepted by: for	Date:
	Instrument loop number

Appendix 6.3.2

Modification control form

Form number:

	System	Loop	Sequence	Display	Trend	Log	Other
Tag/ref number							

Description of problem:

Progression (initials/date as appropriate):

	Initiated	Authorized	Designed	Hazop	Checked	Tested	Implemented
Plant manager							
Works chemist							
Safety advisor							
Project engineer							
Process engineer							
Control engineer							
Instrument engineer							

Documentation:

	Operator's manual	P&ID	Loop diagram	Configuration chart	Sequence flow diagram	Database listing	Discs
Page/ diagram number							
Updated							
Checked							

Appendix 7

Storage tanks – final checks before commissioning

1. Confirm all precommissioning checks complete and outstanding items corrected.
2. Confirm all safety devices tested and ready — for example, relief valves, flametraps, liquid seals, vents and overflows.
3. Where appropriate bund drain is closed and bund free of debris.
4. Tanker offloading connections, hoses and so on are available and clean.
5. Receiving tank is numbered and labelled in accordance with company standard.
6. Tank and connecting lines are drained and clean.
7. Sampling procedures, analytical method and acceptance criteria have been established.
8. Log sheet is available to record all transfers.
9. Level gauge reads zero.
10. Ancillary equipment has been checked, for example, earthing clips, mixer, nitrogen make-up, temperature controller setting, steam pressure to coil, thermometer reading is showing ambient, steam or electric tracing is on to tanker discharge line and remote operation of emergence valves.
11. Safety equipment is in proper location and safety showers work. Material safety data is readily available and Medical Room has treatment information.
12. Containers for collecting sample forerunning and hose drainings are on hand.
13. Equipment and/or material for dealing with spillages is readily available.
14. Operators are familiar with operating procedure, potential hazards and emergency procedure.
15. Confirm delivery quantity, date and time of first tanker.

At the end of the discharge:
1. Check weight received with tanker delivery note.
2. Commission heating/mixing systems as required.
3. Establish inert atmosphere if appropriate.
4. Complete log sheet.
5. Monitor level, temperature and pressure at agreed frequency.
6. Resolve outstanding minor problems.
7. Hand over formally to Production.

Index